73rd Conference
on Glass Problems

73rd Conference on Glass Problems

*A Collection of Papers Presented at the
73rd Conference on Glass Problems
Hilton Cincinnati Netherland Plaza,
Cincinnati, Ohio
October 1–3, 2012*

Edited by
S. K. Sundaram

The
American
Ceramic
Society

WILEY

Published by John Wiley & Sons, Inc., Hoboken, New Jersey.
Published simultaneously in Canada.

For general information on our other products and services or for technical support, please contact our
Customer Care Department within the United States at (800) 762-2974, outside the United States at
(317) 572-3993 or fax (317) 572-4002.

Wiley also publishes its books in a variety of electronic formats. Some content that appears in print may
not be available in electronic formats. For more information about Wiley products, visit our web site at
www.wiley.com.

Library of Congress Cataloging-in-Publication Data is available.

ISBN: 978-1-118-68663-8
ISBN: 978-1-118-68666-9 (special edition)
ISSN: 0196-6219

Printed in the United States of America.

10 9 8 7 6 5 4 3 2 1

Contents

COATINGS AND STRENGTHENING

REFRACTORIES

PROCESS CONTROL AND MODELING

Foreword

The 73rd Glass Problem Conference is organized by the Kazuo Inamori School of Engineering, Alfred University, Alfred, NY 14802 and The Glass Manufacturing Industry Council, Westerville, OH 43082. The Program Director was S. K. Sundaram, Inamori Professor of Materials Science and Engineering, Kazuo Inamori School of Engineering, Alfred University, Alfred, NY 14802. The Conference Director was Robert Weisenburger Lipetz, Executive Director, Glass Manufacturing Industry Council, Westerville, OH 43082. The themes and chairs of six half-day sessions were as follows:

Glass Melting
Glenn Neff, Glass Service, Stuart, FL and Martin Goller, Corning Incorporated, Corning, NY

Melting, Raw Materials, Batching, and Recycling
Phil Tucker, Johns Manville, Denver, CO and Robert Weisenburger Lipetz, GMIC, Westerville, OH

Coatings, Strengthening, and Other Topics
Jack Miles, H. C. Starck, Coldwater, MI and Martin Goller, Corning Incorporated, Corning, NY

Refractories
Matthew Wheeler, RHI US LTD, Batavia, OH and Thomas Dankert, Owens-Illinois, Perrysburg, OH

Warren Curtis, PPG Industries, Pittsburgh, PA and Elmer Sperry, Libbey Glass, Toledo, OH

Process Control and Modeling
Bruno Purnode, Owens Corning Composite Solutions, Granville, OH and Larry McCloskey, Toledo Engineering Company, Toledo, OH

Preface

A tradition of this series of Conference, started in 1934 at the University of Illinois, includes collection and publication of the papers presented in the Conference. The tradition continues! The papers presented at the 73rd Glass Problems Conference (GPC) have been collected and published as the 2012 edition of the collected papers.

The manuscripts included in this volume are reproduced as furnished by the presenting authors, but were reviewed prior to the presentation and submission by the respective session chairs. These chairs are also the members of the GPC Advisory Board. I appreciate all the assistance and support by the Board members. The American Ceramic Society and myself did minor editing of these papers. Neither Alfred University nor GMIC is responsible for the statements and opinions expressed in this volume.

As the incoming Program Director of the GPC, I am truly excited to be a part of this prestigious historic conference series and continue the tradition of publishing this collected papers that will be a chronological record of advancements in the areas of interest to glass industries. I would like to record my sincere appreciation of impressive service of Charles H. Drummond, III as the Director of this Conference from 1976 to 2011. I am thankful to all the presenters at the 73rd GPC and the authors of these papers. The 73rd GPC continues to grow stronger with the support of the audience. I am deeply indebted to the members of Advisory Board, who helped in every step of the way. Their volunteering sprit, generosity, professionalism, and commitment were critical to the high quality technical program at this Conference. I would like to specially thank the Conference Director, Mr. Robert Weisenburger Lipetz, Executive Director of GMIC for his unwavering support and strong leadership through the transition period. I look forward to working with the Advisory Board in the future.

S. K. SUNDARAM
Alfred, NY
December 2012

Acknowledgments

It is a great pleasure to acknowledge the dedicated service, advice, and team spirit of the members of the Glass Problems Conference Advisory Board in planning the entire Conference, inviting key speakers, reviewing technical presentations, chairing technical sessions, and reviewing manuscripts for this publication:

Kenneth Bratton—*Emhart Glass Research Inc. Hartford, CT*
Warren Curtis—*PPG Industries, Inc., Pittsburgh, PA*
Thomas Dankert—*Owens-Illinois, Inc., Perrysburg, OH*
Martin H. Goller—*Corning Incorporated, Corning, NY*
Robert Weisenburger Lipetz—*Glass Manufacturing Industry Council, Westerville, OH*
Larry McCloskey—*Toledo Engineering Co. Inc. (TECO), Toledo, OH*
Jack Miles—*H.C. Stark, Coldwater, MI*
Glenn Neff—*Glass Service USA, Inc., Stuart, FL*
Bruno A. Purnode—*Owens Corning Composite Solutions, Granville, OH*
Elmer Sperry—*Libbey Glass, Toledo, OH*
Phillip J. Tucker—*Johns Manville, Denver, CO*
Mathew Wheeler—*RHI US LTD, Batavia, OH*

Glass Melting

ENERGY RECOVERY FROM WASTE HEAT IN THE GLASS INDUSTRY AND
THERMOCHEMICAL RECUPERATOR

Hans van Limpt and Ruud Beerkens
CelSian Glass and Solar
The Netherlands

ABSTRACT/SUMMARY

Energy is expensive and therefore most glass producers are looking for methods to optimize the energy
efficiency of their glass melting furnaces. For glass furnaces good housekeeping in combination with
energy recovery from waste heat may result in energy savings between 15 and 25%.
Large parts of the energy supplied to fossil-fuel fired glass furnaces are lost through the chimney. Even
efficient regenerative or oxygen-fired furnaces typically show losses of 25-35 % of the total glass
furnace energy input through the stack. Different types of flue gas heat recovery and other energy
saving measures are analysed to show their energy efficiency improvement potential for industrial
glass furnaces.
Different options to re-use the waste gas heat of glass furnaces are:
   o Batch & Cullet Preheating;
   o Drying and preheating of pelletized batch by flue gas heat contents;
   o Application of a Thermo-Chemical Recuperator (TCR) to covert natural gas, water and waste
      heat into a high calorific preheated fuel: syngas or reformer gas;
   o Steam and/or electricity generation by steam or organic vapours (ORC);
   o Natural gas and/or oxygen preheating.

INTRODUCTION

Figure 1 shows a typical distribution of energy supplied to an air-fired regenerative float glass furnace
and figure 2 shows a typical energy balance for a modern end-port fired regenerative container glass
furnace. Note, the relatively large part of energy losses by the flue gases despite the application of
effective regenerator systems for combustion air preheating. Today, batch & cullet preheating systems
[1 -7] are applied in the container glass sector for a few (container) glass furnaces using more than 60
% cullet. Energy savings of 12 to 18 % are reported. A new generation of batch/cullet preheaters, that
can operate with lower cullet fractions is in development.
Also indirect preheating of cullet only with steam is an alternative way to recover waste gas heat.
Other methods shortly described in this paper are: steam generation using flue gas heat contents, batch
pelletizing and pellet preheating, application of a so-called Thermo-Chemical Recuperator (TCR) and
preheating of fuel (natural gas) and/or oxygen.
Preheating of pelletized batch is an alternative for glass furnace without cullet or relatively low cullet
levels. On-site, glass forming raw material batch is pelletized, dried and preheated with heat of the flue
gases and, preheated pellets are introduced into the furnace. Figures 1 and 2 show that waste gas (flue
gas) heat / energy recovery is probably the most important potential for energy savings in glass
melting.

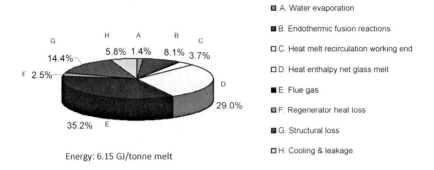

A. Water evaporation
B. Endothermic fusion reactions
C. Heat melt recirculation working end
D. Heat enthalpy net glass melt
E. Flue gas
F. Regenerator heat loss
G. Structural loss
H. Cooling & leakage

Energy: 6.15 GJ/tonne melt

Figure 1   Energy distribution in modern regenerative float glass furnace without
working end, including re-circulating melt (return flow from working-end).

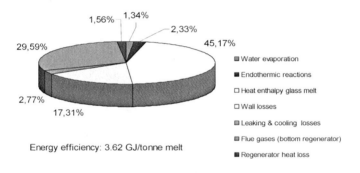

Water evaporation
Endothermic reactions
Heat enthalpy glass melt
Wall losses
Leaking & cooling losses
Flue gases (bottom regenerator)
Regenerator heat loss

Energy efficiency: 3.62 GJ/tonne melt

Figure 2   Energy balance of modern & energy efficient container glass furnace (end-port
regenerative, natural gas - air fired, 84% cullet), no batch preheater.

In the average Dutch glass industry, the glass melting process consumes almost 60 % of the total
(primary) energy input in the primary glass industry. But for some sectors, the melting process
consumes even more than 60 %, such as in float glass production (about 85-90 %) and container glass
production (60-65 %).

SYSTEMS FOR SECONDARY FLUE GAS HEAT RECOVERY

Typical temperatures of flue gases downstream regenerators of glass furnaces are in the range of 450-550 °C. For recuperative furnaces, flue gas temperatures may be above 800 °C or even 900 °C and for oxygen fired furnaces, flue gas temperatures before quenching or dilution may be as high as 1400-1450 °C. A high temperature is advantageous from a thermodynamic point of view, a high temperature difference between a medium that releases energy to a medium at lower temperature level will favour the quantity of energy/heat that can be transferred and the heat transfer rate.

However, at high temperatures, material constraints may limit some methods of heat transfer and flue gas temperatures need to be reduced before secondary heat recovery.

Kobayashi [8] reported about the development of a high temperature batch & cullet preheating system being in development especially for flue gases at very high inlet temperatures e.g. in combination with oxygen-fuel fired glass furnaces.

Production of steam

Production of steam, using flue gas heated boilers (waste heat boilers) and economizers. Today, systems able to handle flue gases of soda-lime-silica glass furnaces have been applied with success, using self-cleaning systems. One of these systems is based on a so-called fire-tube boiler (flue gases flow through pipes/tubes that are bundled in a large vessel filled with water under pressure), producing typically steam between 20-30 bar and 200-250 °C or higher, depending on the flue gas inlet temperature [9], and required steam pressure. Other systems are used as well even for flue gas inlet temperatures above 1200 °C. Such systems provide steam that can be used for several purposes:

a. Preheating cullet [9];
b. Wetting the raw material batch with high temperature humidity to avoid soda clogging below 35.4 °C;
c. Steam to drive turbines that may produce cooling air or pressurized air for the forming machines (e.g. IS machines);
d. Steam to drive turbines connected to an electricity generator, making own power;
e. Steam for heating of buildings;
f. Process steam for other process steps in a glass factory, e.g. preheating of fuel oil;
g. Steam for neighbouring companies, often the transport of heat (in form of steam) requires insulated ducts and investment costs may be high.

Steam may also be supplied to municipalities. However, the supply of this heat/steam cannot be guaranteed in all cases for instance during flue gas system or glass furnace maintenance.

Energy savings by cullet preheating for regenerative container glass
furnace: different cullet % in batch (320 tons glass melt/day)

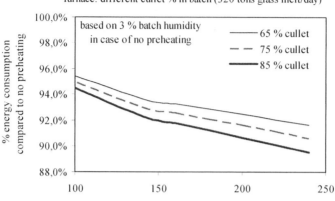

Figure 3   Energy savings for regenerative container glass furnace by cullet preheating at different
             temperature levels (heating medium: steam)

In most cases option 1 and sub-options a-g depend very strongly on economic and ROI (return on
investment) considerations. Especially options 1c – 1d may face high investment costs and longer time
periods for their Return On Investment, but at the end these methods may be cost-effective on the
longer term. Electricity generated from steam can directly be used for oxygen generation or for electric
boosting or be supplied to the electricity grid. Examples are shown in the German glass industry.

Self cleaning Steam boiler vessel

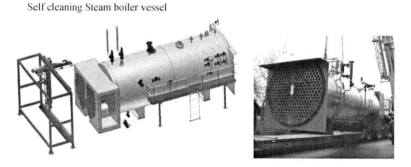

Figure 4   Waste heat boiler for steam generation using glass furnace flue gases as supplied by
             OPTIMUM [9].

The application of steam for cullet drying and preheating has been discussed in [9]. Steam can be applied to preheat cullet up to 200-220 °C. For furnaces using 75 % cullet in the batch energy savings of 8-10% seem possible for air-fired regenerative (see figure 3) and even ±10-12 % for oxygen fired furnaces (with additionally 10-12 % oxygen savings).

Although, the technology of steam production from flue gas heat from glass furnaces is a mature technique and fouling plus corrosion problems can be handled and minimized, demand for steam is not always present or steam distribution over longer distances may be too expensive. For flue gas temperatures of 500 °C, about 50-60 % of the sensible energy of the flue gases (reference temperature is 0 °C) can be converted in the energy contents of the steam. In case of 1000 °C flue gas temperature inlet, (oxygen-fired glass furnace situation), the steam generation can be more efficient, 70-80 % of the sensible heat can be recovered. Figure 4 shows a steam boiler, with tubes heated by flue gases flowing through these tubes. This combination of a special designed fire tube boiler with an in-line cleaning system has been successfully applied in multiple float and container glass plants.

Batch and cullet preheating
The preheating of batch with more than 50-60% cullet, is considered as a mature technology in the glass industry and exploited since the mid 1980-ties, especially in the container glass industry in Germany. In case of complete batch (including cullet) preheating, typically at temperatures of 250-325 °C, the energy savings on a fossil-fuel fired glass furnace are in the range between 12-18 % (or sometimes more in connection with increased melting capacity). Highest energy cost savings can be achieved by increasing the pull on a furnace when preheating batch without increasing fuel input or by keeping constant pull and fuel input but lowering electric boosting. Probably the pull can be increased by 18 to 20 % if batch and cullet would be preheated to 290 °C (reference: batch and cullet at room temperature with 3 % moisture).

Despite the high capital costs, pay-back times low 2-3 years are reported, of course strongly determined by energy prices. Today lifetimes up to 20 years are expected for batch-cullet preheating systems [1-5]. The batch preheat temperature is very important for the energy consumption of the glass furnace. In case of high batch humidity and for instance cullet with high water content, the preheat temperatures are limited due to the evaporation heat required to evaporate the water.

One of the most important aspects of concern is charging a completely dry preheated batch into a glass furnace. This, may lead to dusting in the doghouse area but can also increase batch carry-over into the flue gases. Special doghouse designs are in development to avoid dusting problems in the charging area and limit the dust release from the batch blanket when entering the melting furnace [10].

There are direct contact preheaters [1, 5] (direct contact between flue gas and batch/cullet) and in-direct preheaters [2, 3, 4] on the market. The direct contact preheaters offer the possibility to act as a scrubber device: raw materials in the batch absorb acid species (SO_2, SO_3, HF, HCl, SeO_2) from the flue gases. But direct preheaters often show high dust loads in the flue gases downstream the preheater system. This asks for a high efficient filtering (ESP or bag filter) system.

Today, a few manufacturers of such batch preheating systems are developing adapted configurations to enable the preheating of batch with low cullet levels, down to 10 % for instance, without clogging problems [3, 4, 7, 17, 18]. Figure shows a sketch of the newly developed Zippe preheater [18].

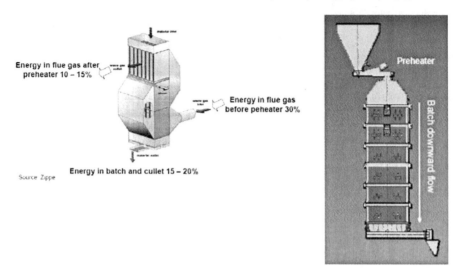

Fiigure 5   Sketches of a Zippe batch preheater (left) [18] and SORG batch preheater (right) [17]

Main issue is the release of water and the removal of the vapours from the batch and cullet and the handling of very dry preheated batch during transport to and into the glass furnace. The SORG batch preheater concept uses vibrational movements to prevent clogging and plugging [17], while Zippe the problem of 'clumping or clogging' solved by an improved evaporation section [18].

Drying an preheating of pelletized batch

Increasing energy prices and CO_2 prices will stimulate the application of batch preheating systems in glass industry, today it is limited applied in the container glass industry. Application of batch preheaters in other sectors of the glass industry probably requires preheaters for batch with low cullet levels [17-19] or for very fine batch materials (that may need pelletizing prior to preheating).

Pelletizing the batch, using water and binder(s) and drying and preheating the formed pellets (granules made of the raw material mixture) by flue gas heat. This concept is still in development. Typical sizes of the agglomerated batch granules are 3-15 mm. The advantage is that transport and charging of such pellets (assuming that pellets are stable and do not break or form dust fragments) is relatively easy and not prone to carry-over or dusting. Another advantage is that the melting process is accelerated by using pelletized batch [11], because of the intensive and close contact between the different raw material species that react and fuse upon heating in the glass furnace. Each pellet contains all batch ingredients at the correct proportions and this will result in a very homogeneous melt formed from the fusion process of these pellets. Depending on the batch and glass composition a certain quantity of binding agents (water glass, caustic soda, etcetera) may be required or relatively high levels of water have to be added. Often, the grain sizes of the raw materials have to be adapted for making the batch suitable for compaction/pelletizing. High water contents of the green pellets will demand relatively high quantities of energy (from the flue gases) upon drying and consequently will reduce the batch preheat temperature. Depending on the flue gas temperature level, ratio of flue gas volume flow versus batch pull, water content of the pellets, preheat temperatures of 175-300 °C can be realized for regenerative glass furnaces. Figure 6 shows the effect of the water content of the pellets on the achievable pellet preheat temperature for a typical container glass furnace (end-port regenerative).

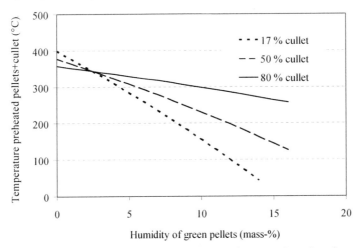

Figure 6   Preheating of pelletized batch by flue gases from container glass furnace (end-port regenerative), depending on water content in pellets. At high water content, flue gas energy contents are mainly used for water evaporation.

The equipment includes a pelletizing disc (rotating pan), a pellet drying plus pre-heating system, pellet transport and charging system into the glass furnace, sieves that separate dust from the pellets and to be reprocessed, the flue gas system includes a filter downstream the pellet drying and preheating unit(s). This technology is still in development, but offers energy savings up to 15-20 %, improved glass homogeneity and expected increased pull rates (15 – 25 %) on existing glass furnaces.

Thermo-Chemical Recuperator
Thermo-Chemical Recuperator (TCR): In a TCR system, natural gas is converted into <u>hot</u> synthesis gas (mainly CO and hydrogen) with higher energy content than this natural gas, see the scheme below. For this highly endothermic reaction, waste gas heat is used. This technology is interesting for oxy-fuel as well as recuperative furnaces. Energy savings can be 25 % or higher, depending on the temperature

of the temperature of the reformer and obtained reformer gas, and depending on the type of furnace. Highest energy savings are expected for recuperative furnaces. First flue gas heat is used to generate steam, steam and de-sulfurized natural gas are mixed and further heated, and finally the natural gas (mainly CH_4) and steam (H_2O) mixture is exposed at 600-900 °C to a catalyst.

Principle of TCR-system

- Production of hot synthesis gas (CO, H_2)
- Required heat is recovered from the waste gases
- Combustion of synthesis gas (CO, H_2) in the glass furnace

	Heat $CH_4 + H_2O \longrightarrow CO + 3 H_2$	
	25 °C	700 °C
Moles of gas	1	4
Net calorific value (MJ/kmol)	802	283 242
Heat of combustion (MJ)	802	1090
Enthalpy (MJ)	≈ 0	83
Total heat content (MJ)	802	1173

The hydrocarbon molecules and water react and form hydrogen and CO, so-called syngas (reformer gas). The hot syngas (rich in hydrogen) can be used as fuel for the glass melting process. The syngas composition depends on the molar ratio of water (steam) to hydrocarbon molecules, the temperature of the reformer and the catalyst activity [13]. This TCR-system requires: a steam boiler (heated by flue gas), super heater, a natural gas de-sulfurization process, a recuperator (in case of an air-fired furnace), a reformer system using a catalyst. In the past often nickel catalyst were applied. These catalysts are relatively cheap but sensitive to sulfur compounds and to coke formation. Nowadays precious metal catalyst are used that are less sensitive to sulphur and to coke formation. These catalysts are more expensive but only a small percentage of the total equipment cost.

Furthermore, special burners seem to be required for firing this hot syngas in the glass furnace. Thermo-Chemical Recuperators are probably most suited in combination with recuperative or oxygen-fired furnaces. It requires high flue gas temperatures (> 800 °C). Constraints or barriers today for application of a TCR are: no experiences in glass industry, system is not familiar to glass producers, required capital for investments, and the development for special burners and reformer systems are required. Fouling of the reformer system (which converts the natural gas and steam into syngas by a catalyst and using heat from flue gases) by condensation products of the glass furnace flue gases may be an aspect of concern.

Figure 7 shows a diagram and photo of a TCR system as developed by the company HyGear [13].

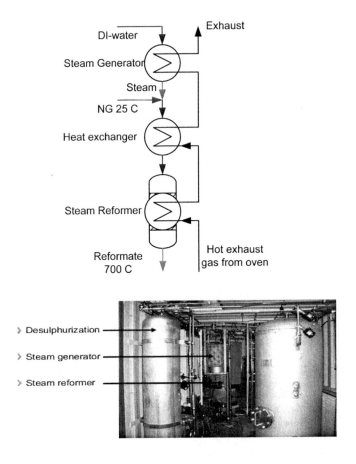

Figure 7 Diagram and photo of process steps in the Thermo-Chemical Recuperator process [13]. Temperatures given in °C.

It is possible to apply a TCR system for supplying part of the energy input to the furnace via reformer gas instead of natural gas and using a part of the flue gases to heat the reformer and to generate the steam. Such a pilot plant at an existing glass furnace could be a first step to test existing TCR technology at an existing glass furnace and measure the effect on energy consumption, emissions, burner behaviour, catalyst and reformer behaviour and fouling aspects in a test period of 6 months to 2 years.

Preheating of natural gas and/or oxygen

Natural gas preheating by flue gasheat has been applied in a limited number of cases in the glass industry since many years (e.g. in Germany, Czech Republic). Natural gas preheating up to about 350-400 °C has been practised with energy savings due to this preheating in the order of 3 %.

Recently, a system has been demonstrated at an oxygen-fired float glass furnace comprising a recuperator to cool flue gases of the oxygen-fired glass furnace and preheating air to, for instance, 700-800 °C (see figure 8).. The hot air (< 800 °C) from this recuperator system is used to preheat natural gas and oxygen for firing the float glass furnace. Oxygen preheat temperatures of 550 °C and natural gas preheating up to 450 °C is reported [12, 14]. The preheating of natural gas and oxygen in this way leads to considerable gas and oxygen savings in this furnace. Energy and oxygen savings are about 8 % (calculated from energy balance model). There is still energy left in the flue gases for steam generation.

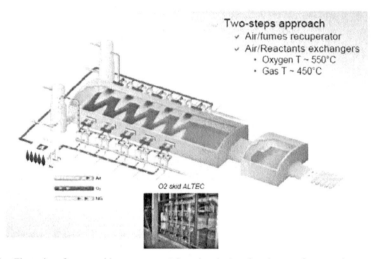

Figure 8   Float glass furnace with recuperator (air preheating) and preheaters for natural gas and oxygen (Air Liquide/AGC) [14].

ORC Electricity generation

Electricity generation from waste heat, especially from lower temperature waste heat is possible with the so-called Organic Rankine Cycle process [15]. The waste heat from many processes in industry offers a huge potential for energy recovery. The Rankine cycle using organic vapours instead of steam expansion to drive a turbine offers the possibility to use lower temperature levels for generating the high pressure vapour. The working range of an ORC medium (organic) is between 50-300 °C. The efficiency of the ORC cycle depends on the selected organic fluid, the heat transfer between the waste heat and the organic fluid to be evaporated, the expander, the generator and so forth. The real efficiency of an ORC process is less than 50 % of the Carnot energy release.

The ORC organic fluid (replacing steam) can be heated by low temperature waste heat coming from furnaces exhausts, feeder walls, annealing lehr waste heat, etcetera.

An example: from flue gases at a flow of 100.000 Nm^3/hr (large float glass furnace), cooling down from 260 to 160 °C could provide a thermal power of 2.1 MW.

Possible applications for ORC electricity generation in glass industry: flue gases from feeders, refiners, flue gases from glass furnace after filter, hot gases from driers or étuves (glass wool), gases from annealing lehrs or tin bath and heat 'the collected from feeders/forehearths' heat-releasing surfaces. There are for instance standard ORC Units for a capacity of 50 till 250 kW electricity.. Typical liquid in the system: Honeywell r245fa or Solkatherm [15], typically heated by a medium of 80-150 °C.

Other methods

Other methods for the reclaim of sensible heat contents of glass furnace flue gases, may be in development, but are hardly or not applied at the moment. Additional combustion air preheating (additional to recuperator or regenerator) can be considered, but generally this is not very effective anymore in case of a modern regenerative glass furnace, with regenerator efficiencies approaching the thermodynamic limit of about 75 %. Use of flue gas heat or steam for an absorption cooling system may be considered where a cooling medium is needed. Energy balance models [6] for glass furnaces including regenerators, recuperators and secondary waste heat recovery systems are available to determine accurately the effect of secondary heat reclaim on CO_2 emissions and energy consumption (total and specific) of glass melting processes and savings in energy, oxygen and CO_2 costs. Such a model supports the selection of the most appropriate cost-effective solution, which depends on type of glass furnace, local energy costs, cullet % in batch, size of production and space availability [16].

CONCLUSIONS

Secondary heat recovery from flue gases of glass furnaces is only applied in a limited number of cases. The further exploitation of the heat contents of flue gases of oxygen fired furnaces or flue gases exiting regenerators or recuperators often requires additional investment costs or techniques that are not considered as main activities in glass manufacturing. Mature technologies today for secondary heat recovery are batch preheating (in case of batches > 50 % cullet), steam generation and natural gas preheating. Promising technologies that need further developments and maybe to be further developed as cheaper installations: batch pelletizing in combination with pellet drying & preheating by glass furnace flue gases and the Thermo-Chemical-Recuperator (TCR) technology. Batch preheating for batches with low fraction of cullet are recently (2011) in development and will soon be applied on a larger scale. Therefore, existing batch preheat concepts have been modified and furnace designs have to be adapted to avoid too much carry-over and dust formation when charging dry batches into a glass melting tank furnace [17, 18, 19]

Main concerns today are:
fouling and batch carry-over problems encountered in cases of charging very dry preheated batch into glass furnaces;

capital costs for preheating systems and electricity generation from waste heat;
preheating of batches with very low cullet content;
feasibility of pelletizing and feasibility and demonstration of Thermo-Chemical Recuperator technologies.
Obtaining experiences with ORC and TCR in the glass industry.

Most developments have been applied in the container glass sector (batch preheating, natural gas preheating) and float glass sector (steam generation, electricity generation, oxygen and natural gas preheating), some technologies may be applicable also in other sectors of the glass industry.
Especially, in case of the final development and proof of the new generation batch preheaters that allow preheating of cullet-lean batches, without contaminating the batch by abrasion of the inner steel construction, application of batch preheating becomes possible for non-container glass furnaces as well.
Many methods can reclaim 40-70 % of the flue gas heat contents, depending on the flue gas inlet temperature. This may result in glass melt process energy savings of 12 to more than 20 %.

LITERATURE

[1] H.J Barklage-Hilgefort: 3 Jahre Betriebserfahrung mit einer querbeheizte Regenerativ-wanne mit Gemengevorwärmung. Vortrag vor dem Fachausschuß II der
DGG, 22. October 1998
[2] B.H. Zippe: Reliable batch and cullet preheater for glass furnaces. Glass Technol.
35 (1994) no. 2, pp. 58-60
[3] P. Zippe: Recent Developments of Batch and Cullet Preheating in Europe – Practical Experiences and Implications. Proceedings of 71st Conference on Glass Problems, October 19-20, 2010 Columbus Ohio, Am. Ceram Soc. Ed. Charles Drummond, III, pp. 3-18
[4] G. Lubitz: Praxiserfahrungen mit einer neuen Generation von Schmelzgutvorwär-mern. 85. Glastechnische Tagung, Saarbrücken, 30. Mai-1. Juni 2011, Kurzfas-sungen. pp. 40-41
[5] J. Herzog; R.J. Settimo: Cullet preheating: the realistic solution for all glass
furnaces with cullet addition. Ceram. Eng. Sci. Proc. 13 (1992) no. [3-4], pp. 82-90
[6] R. Beerkens: Energy Saving Options for Glass Furnaces & Recovery of Heat from their Flue Gases - And Experiences with Batch & Cullet Pre-heaters Applied in the Glass Industry. 69th Conference on Glass Problems, Columbus Ohio, USA 3. & 4. November 2008, pp. 143-162 Ed. Charles Drummond III, Am. Ceram. Soc. (2009) Published by John Wiley & Sons, Inc. Publication
[7] M. Lindig: New Sorg Batch Preheating System. GMIC GlassTrend Workshop: Waste Heat Management in the Glass Industry, October 21, 2010, Columbus OH.
[8] S. Kobayashi; E. Evenson; E. Miclo: Development of an Advanced Batch/Cullet
Preheater for Oxy-Fuel Fired Glass Furnaces. Glass Trend workshop on Energy Efficiency and Environmental Aspects of Industrial Glass Melting, 10. - 11. May 2007 Versailles, France
[9] N. Rozendaal; R. Beerkens, A. Habraken: Cullet Preheating by Steam Generated
from Flue gas heat of glass furnaces. GMIC GlassTrend Workshop: Waste Heat Management in the Glass Industry, October 21, 2010, Columbus OH.

[10] M. Lindig: Das Integrierte Konzept zur Gemengebehandlung am Ofen. 85. Glastech. Tagung, Saarbrücken, 30. Mai-1. Juni 2011, Kurzfassungen. pp. 41-42

[11] H. Van Limpt; E. Engelaar; R. Beerkens: Melting kinetics of normal glass forming raw material batches and pretreated batch. Proceedings/Program and Book of Brazil. Pp. 34, paper 0013

[12] Y. Joumani; L. Jarry; J.-F. Simon; A. Contino; O. Douxchamps: Results of a technology recovering waste heat to preheat oxygen and natural gas for oxy-fuel furnaces. GMIC GlassTrend Workshop: Waste Heat Management in the Glass Industry, October 21, 2010, Columbus OH.

[13] A. Scholten, H. Van Limpt: Feasibility of Thermo-Chemical Recuperator TCR for Glass Industry. NCNG workshop: Energy Recovery from Waste Heat in the Glass Industry. 6. October 2011, Eindhoven.

[14] Y. Joumani; F. Bioul; A. Contino; J-F. Simon; L. Jarry; B. Leroux; O. Douxchamps; J. Behen: First oxygen fired float glass furnaces equipped and operated with a new heat recovery technology. Energy Recovery from Waste Heat in the Glass Industry. 6. October 2011, Eindhoven.

[15] G. De Graeve: Electricity from waste heat with Organic Rankine Cycle. NCNG Workshop: Energy Recovery from Waste Heat in the Glass Industry. 6. October 2011, Eindhoven.

[16] H. Van Limpt: Calculation Tool for Selection of Energy Recovery from Flue Gases from Glass furnaces, Agentschap NL/TNO/NCNG presented at: NCNG Workshop: Energy Recovery from Waste Heat in the Glass Industry. 6. October 2011, Eindhoven.

[17] M. Lindig: New developments in batch preheating. NCNG Workshop: Energy Recovery from Waste Heat in the Glass Industry. 6. October 2011, Eindhoven.

[18] P. Zippe; G. Lubitz: Practical Experiences with a New Generation of Batch and Cullet Preheater. NCNG Workshop: Energy Recovery from Waste Heat in the Glass Industry. 6. October 2011, Eindhoven.

[19] S.R. Kahl: Recent developments in Batch preheating. NCNG Workshop: Energy Recovery from Waste Heat in the Glass Industry. 6. October 2011, Eindhoven.

[20] R. Beerkens; H. Van Limpt: Overview of Heat Losses by Flue Gases of Glass Furnaces & Existing Heat Recovery Methods. NCNG Workshop: Energy Recovery from Waste Heat in the Glass Industry. 6. October 2011, Eindhoven.

[21] J. Schep; E. van Leeuwen: Waste Heat Recovery Studies within O-I. NCNG Workshop: Energy Recovery from Waste Heat in the Glass Industry. 6. October 2011, Eindhoven.

[22] N. Rozendaal: Steam generation & Application for Steam. NCNG Workshop: Energy Recovery from Waste Heat in the Glass Industry. 6. October 2011, Eindhoven.

DEVELOPMENT OF DFC (DIRECT FUSION COMBUSTION) USING OXY-FUEL BURNER

Takayuki Fujimoto
Taiyo Nippon Sanso Co.
Hokuto Yamanashi, Japan.

ABSTRACT

We have developed a new high efficient DFC (Direct Fusion Combustion) oxy-fuel burner to melt powdered material. The powdered material is directly supplied into the flame of DFC burner which is located at the top of a furnace. . In this study, we established the DFC burner's structure and scale-up rules by numerical simulation and actual experiments. We designed a pilot-scale DFC burner (glass feed rate: 420 kg/h) and carried out pilot-scale experiment using glass cullet ($d_{50} \sim 300$ mm). The melting efficiency was achieved 65 % on keeping the temperature of the molten glass over 1300 °C.

INTRODUCTION

Oxy-fuel combustion technology has supported for varied industries including glass manufacturing industry. A new glass melting process employing a new oxy-fuel combustion technology is currently investigated mainly in Japan. It is "In-flight glass melting process (IFM process)".[1,2] A Japanese national project (NEDO) for a jointly development of the IFM process was started by 2005. The IFM process is expected to reduce energy consumption by more than 50 %. In the IFM process, an oxy-fuel burner is installed in the upper part of a vertical furnace. In the furnace, raw material supplied into the burner flame and melted during its fall. The oxy-fuel technology to melt the raw material would play a central role in the IFM process.

We have improved oxy-fuel technologies to heat and melt powdered material for about 20 years. As an example, we developed CERAMELT[TM], which is a process for making spherical ceramic particles using an oxy-fuel burner.[3-6] In the CERAMELT, the powdered raw material passes through the burner flame and becomes spherical particles because of their own surface tension. The spherical ceramic particles are mainly utilized in the market of semiconductor. Spherical silica particles, for example, are used as filler particles of encapsulating material in semiconductor chips. Another example, we developed FLASH (FLy ASH) melting burner, which is an oxy-fuel burner for melting fly ash of the garbage incineration.[7] The FLASH melting burner is installed in a system for the volume reduction of the fly ash. The fly ash is also supplied into the oxy-fuel burner and melted by the flame. The molten fly ash holds in the downside of the furnace. The molten slag is flowed out continuously and crashed. The crashed slag is used as concrete aggregates.

These conventional burners are designed for the powdered material of which the size is under several tens micro-meter. The new burner, however, should melt particles of several hundreds micro-meter size in the IFM process. Therefore, we started an originally development of a new high efficient oxy-fuel burner named "DFC (Direct Fusion Combustion) burner" to melt the relatively large powder.

STRUCTURE

The basic structures of the DFC burner and the FLASH melting burner are shown in figure 1. The FLASH melting burner has the structure in which the outlet of the powdered material is placed annularly. The structure makes the high dispersibility of supplied powder in the flame.

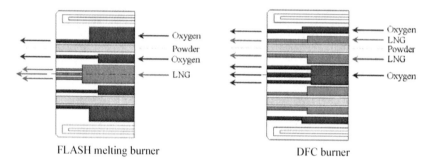

FLASH melting burner   DFC burner

Figure 1: Schematic of conventional FLASH melting burner and DFC burner

We designed the DFC burner based on the FLASH melting burner to improve the heat transfer to the particles in the flame and the heat transfer to the surface of the molten glass. The DFC burner has the structure in which the outlet of the powdered material is also placed annularly and placed between outlets of the fuel. The new structure causes the flow of raw material caught between inside flame and outside flame. Because the flames heat the raw material immediately after the injection, the residence time of the particles in the flame is extended. Therefore, the raw material is heated high effectively during its fall.

NUMERICAL SIMULATION

We examined the DFC burner's structure by a numerical simulation technology, which can investigate particles behavior in oxy-fuel combustion flame [5], with FLUENT 6.3 on bench-scale. The computational models and the conditions are shown in table 1 and 2. We also simulated the FLASH melting burner for the comparison. The results are shown in figure 2 and 3.

Table 1: Computational models

Turbulence	Standard k-ε model [8,9]
Combustion	Probability Density Function model (PDF) [10]
Radiation heat transfer	Discrete Ordinate model (DO) [11]
Multiphase flows	Euler-Lagrangian model

Table 2: Computational conditions

Furnace size		750×750×900 mm	
LNG flow rate		$(m^3(normal)/h)$	11.0
O_2 flow rate		$(m^3(normal)/h)$	26.6
Glass powder feed rate		(kg/h)	145.0
Glass powder	Diameter	(μm)	300
	Density	(kg/m^3)	2500
	Specific heat	$(kJ/(kg \cdot K))$	1.2

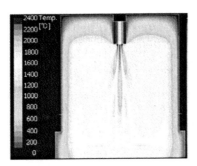
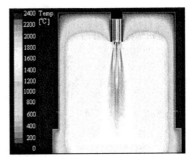

FLASH melting burner   DFC burner

Figure 2: Calculating temperature distribution

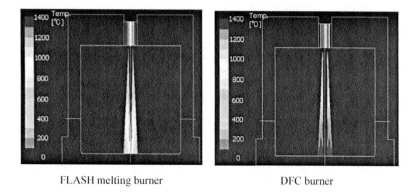

FLASH melting burner   DFC burner

Figure 3: Calculating particle trajectory and temperature

In figure 2, the FLASH melting burner forms the cold areas in the center of the flame. The particles temperature is around 1100 °C on arriving at the surface of the molten glass. Furthermore, the FLASH melting burner makes a cold spot in the center of the molten glass. On the other hand, the DFC burner forms the high temperature and large area in the furnace. Figure 2 shows that the inside flame and the outside flame catch the particle trajectory. The particles temperature is 1380 °C on arriving at the surface of the melt. In figure 3, the particle trajectory of the DFC burner is narrower than one of the FLASH melting burner. The two flames of the DFC burner catch the trajectory of particles and prevent the particles from spreading. We optimized and established the structure of the DFC burner.

BENCH-SCALE EXPERIMENT

Equipment

We actually manufactured the optimized DFC burner for bench-scale and built equipment for the performance evaluation of the burner. Each specification of the equipment is shown in table 3. Figure 4 shows the schematic of the equipment.

Table 3: Specifications of bench-scale equipment

Furnace size	750×750×900 mm
Melt capacity	100 L (250 kg-glass)
Heat input	126 kW
15 temperature sensors for molten glass.	

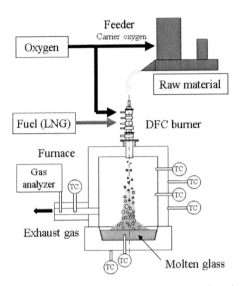

Figure 4: Schematic of the equipment for bench-scale

Evaluation

We evaluated the heat transfer performance of the burners by the melting efficiency η (%) defined in equation (1).

$$\eta = \frac{Q_a}{Q_i} \times 100 \qquad (1)$$

Q_a: Sensible heat of molten glass (kW), Q_i: Heat input (kW)

We chose soda-lime glass cullet powder as raw material, because we didn't need consider the chemical reaction. Then Q_a can be expressed as equation (2).

$$Q_a = \frac{d}{dt}(CWT) = C\left(T \cdot \frac{dW}{dt} + W \cdot \frac{dT}{dt}\right) \qquad (2)$$

C: Specific heat (kJ/kg·°C), W: Melt weight (kg)
T: Temperature difference between melt and air (°C)

On thermal balance, the time change of the temperature difference T is zero. The time change of melt weight W means the feed rate of the raw material. Then the melting efficiency η can be described as equation (3).

$$\eta = \frac{\Delta W}{Q_i} \cdot CT \times \frac{100}{3600} \tag{3}$$

ΔW: Feed rate of raw material (kg/h)

Experiment and Results

The experimental conditions are shown in table 4. Table 5 shows the results of experiment. The melting efficiency of the FLASH melting burner is 43 %. One of the DFC burner is 54%. The temperature of the melt using DFC burner is over 1410 °C. We confirmed that the DFC burner is relatively efficient as compared with the FLASH melting burners. The d_{50} of glass cullet is about 300 μm.

Table 4: Experimental conditions on bench-scale

LNG flow rate	$(m^3(normal)/h)$	11.0
Oxygen flow rate	$(m^3(normal)/h)$	26.6
Oxygen ratio	(-)	1.05
Glass powder feed rate	(kg/h)	125~145

Table 5 Experimental results on bench-scale

Burner type	FLASH melting burner	DFC burner
Melting efficiency (%)	43	54
Temperature of molten glass (°C)	1382	1416

We carried out the experiments that changed the distance between the DFC burner and the surface of molten glass. Figure 5 shows the result. When the distance is around 1m, the melting efficiency is the highest.

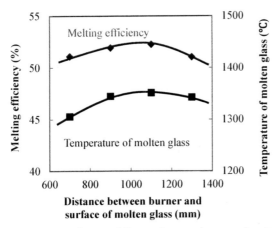

Figure 5: Influence of distance between burner and melt

It is thought that the long residence time of particles causes high heat transfer from flame in the furnace. The result, however, suggests that the heat transfer to falling particles achieved upper limit, when the distance is longer than the suitable one. The distance between burner and the surface of the molten glass causes the decrease of the heat transfer to surface of molten glass. The result shows that it is an important factor for arrangement of a burner to improve melting performance.

SCALE-UP AND PILOT-SCALE EXPERIMENT
Scale-up

We established the scale-up rules to design DFC burners, mainly using numerical simulation. We optimized the scale-up rules for the ejecting velocities and their nozzles' arrangement of the raw material and each gas. We also optimized the rules for the distance between the burner and the surface of the molten glass. On the bases of these optimized rules for scale-up, we designed and manufactured a pilot-scale DFC burner.

Pilot-scale experiment

We constructed pilot-scale equipment in order to confirm the performance and the scale-up rules of the DFC burner. The specifications of the equipment are shown in table 6. Figure 6 shows the picture of the main parts of the equipment.

Table 6: Specifications of pilot-scale equipment

Furnace size	1050×1050×1260 mm
Melt capacity	380 L (950 kg-glass)
Heat input	310 kW

28 temperature sensors for molten glass.

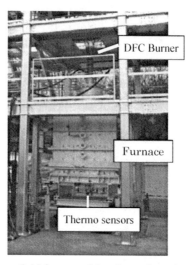

Figure 6: Main parts of the pilot-scale equipment

We carried out the pilot-scale experiment. The experimental conditions are shown in table 7. They are about 3 times of bench-scale experiment. The result is shown in figure 7. The melting efficiency achieved 65%.

Table 7: Experimental conditions on pilot-scale

LNG flow rate	(m^3(normal)/h)	26.5
Oxygen flow rate	(m^3(normal)/h))	64.0
Oxygen rate	(-)	1.05
Glass powder feed rate	(kg/h)	420

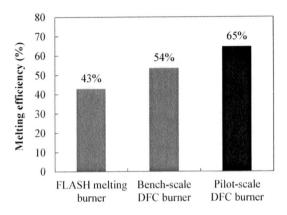

Figure 7: Results of experiments

Verification of Scale-up

We compared the results of numerical simulation and actual experiments with their melt efficiency. Figure 8 shows the comparison. The results of numerical simulation and the actual experiment were corresponded very well. We consider that the scale-up rules are established and it is available for the commercial scale. The calculated melting efficiency is about 70% on the scale.

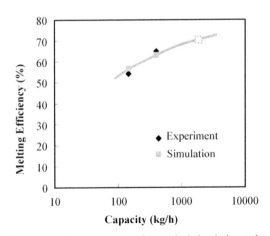

Figure 8: Comparison between results of numerical simulation and experiment

CONCLUSION

We developed a new oxy-fuel burner "DFC burner" which can melt powdered raw material efficiently.

1. DFC burner's structure: The structure in which the raw material outlet is located between inside flame and outside one is most suitable.

2. High efficiency: The melting efficiency was achieved 65 % on pilot-scale, of which feed rate of glass is 420 kg/h.

3. Scale-up rules: The scale-up rules established in this study have enabled the designing of commercial scale DFC burner. The scale is 2 ton-glass/h which equal to 50 ton/day

For the future prospect, we would improve the DFC burner for the continuous flow process and apply to heat other powdered material as well as the glass raw material.

REFERENCES

1) Sakamoto O. Innovative Energy-Saving Glass Melting Technology. Res. Reports Asahi Glass Co., Ltd., 2009, 59, 55-60.

2) Sakamoto O.; Iseda T. Energy Conservation Technology for the Glass Melting Process. J. HTSJ, 2010, 49(207), 14-18.

3) Yajima, T.; Murakami, S.; Miyake S. Manufacturing Technique to the Spherical Inorganic Oxide Particles with the Oxy-Fuel Combustion Flame Method. NIPPON SANSO Technical Report. 1998, 17, 43-53.

4) Miyake, S.; Kinomura, N.; Suzuki, T.; Suwa, T. Fabrication of spherical magnetite particles by the flame fusion method. J. Mater. Sci. 1999, 34, 2921-2928.

5) Yamamoto, Y.; Hagihara, Y.; Kitamura, Y. Investigation of Particle Behavior in the Oxy-Fuel Combustion Flame Using Numerical Simulation. TAIYO NIPPON SANSO Technical Report. 2008, 27, 6-11.

6) Murakami, S.; Suzuki, K.; Hagihara Y. New CERAMELT™ for High-Melting Point Material. TAIYO NIPPON SANSO Technical Report. 2009, 34-35

7) Yamada, S.; Kindo, K.; Miyake S. Development of a town-gas type fly ash melting oxygen burner. Kogyo-Kanetsu (Industrial Heating). 2003, 40(2), 23-28.

8) Ralthby, G. D.; Chui, E. H. A finite-volume method for predicting a radiant heat transfer in enclosures with participating media. J. Heat Transf. 1990, 112(2), 415-423.

9) Chui, E. H.; Ralthby, G. D. Computation of radiant heat transfer on a nonorthogonal mesh using the finite-volume method. Numer. Heat Transf. PartB. 1993, 23(3), 269-288.

10) Sivathanu, Y. R.; Faeth, G. M. Generalized state relationships for scalar properties in nonpremixed hydrocarbon/air flames. Combust. Flame. 1990, 82(2), 211-230.

11) Jones, W. P.; Launder, B. E. The prediction of laminarization with a two-equation model of turbulence. Int. J. Heat Mass Transf. 1972, 15(1), 301-306.

FINDING OPTIMAL GLASS COMPOSITIONS

Oleg A. Prokhorenko
Laboratory of Glass Properties International, LLC
Farmington Hills, MI, USA

Sergei O. Prokhorenko
Ecole Centrale,
Châtenay-Malabry, France

ABSTRACT
Development of a new glass composition includes formulation of trial glasses, melting, and testing their physical properties till the target composition is found. Often, the authors use additive schemes for calculation of properties by composition. Compositional search is, usually, based on systematic study of properties as functions of composition, which is an iterative process. A number of glasses in each series, and a number of iterations depend on amount of available literature data, and required degree of improvement over the known analogues. The present paper describes a method of development of new glasses, which uses the following three elements in compositional search together: reliable structure-properties correlations within wide compositional ranges; multidimensional optimization of performance and cost; combinations express and precision melting and testing methods. All three elements are used to allow one step-wise systematically narrow the range of search to find target glass composition(s) within shorter period of time.

INTRODUCTION
At present the glass industry has well-established range of glass formulas for manufacturing float glass, glass containers, glass fibers, and many other products. Many of them are produced with minimal or no modifications for decades. The reasons are as follows. Regimes of glass melting, and processing were adjusted as a result of long trials-and-errors period. Types and sources of raw materials were chosen. Properties of the final products were balanced to some optimum. Successful modification of the existing glass formulas or finding replacements characterized by the best possible (for a given glass system) cost / performance combinations is, on our opinion, based on three conditions. One – is availability of mathematical algorithms, which allow finding global minimums or maximums of key physical properties, (melting, forming and annealing/tempering range temperatures, density d(T), dielectric constant and loss tangent, $\Delta T_{3\text{-liq}} = T_3 - T_{liq}$, tensile strength, tensile modulus and others), and cost as a result of multi-dimensional optimization. Two – is availability of simple, and reliable correlations between properties of glasses or melts and their composition, capable in calculating changes of crucial physical properties caused by relatively small compositional variations. Such correlations or functions are essential part of optimization process. Three – is a choice of experimental methods ranging from express melting and testing of numerous glasses within a wide compositional range to fine melting and high precision measurements for final adjustment of the final glass compositions for further production. Using this method more trial glass compositions are melted, and tested within the shorter period of time.

This, basically, iterative process save valuable resources: cost and time. Having composition optimization software with a set of functions determining key physical properties available glass technologists can continue optimization process further to generate additional high-performance glass formulas.

The present paper considers all three components of the methodology in detail. It describes the methods of calculation of key glass properties by their composition, the algorithm of multidimensional optimization on cost-performance base, and different techniques of melting and testing of experimental glasses, which combined allow to deliver the final high performance low cost glass formulas.

Here we should note that the present work is made possible due to intensive long-term of LGP research team supported by our colleagues from US, European, and Asian glass companies and universities.

There is a connection between volume of production of certain type of commercial glass and changes of current glass composition practitioners are ready to go for. Such high-tonnage segments as float glass, container glass, and conventional fiberglass industries are more conservative to make quick changes due to many known reasons. Production of many specialty glasses such as thin-sheet LCD/LED display glass, high-performance fiberglass, low-melting sealing glass, and many others requires continuous innovations to offer competitive products to the market. In order to support existing demand for improved specialty glass compositions, and be prepared for coming need to optimize conventional glass formulas one should develop and validate a strong methodology. Glass industry accepts new glass compositions in case they bring significant benefits, and allow making as little as possible changes to the existing technological regimes. Thus, developing new glass composition one should accomplish dual task: to deliver a new composition providing better consumer characteristics, and having key technological parameters compatible with existing process. Reduced energy consumption, cost and waste, and environmental safety are considered to be critical add-ons.

OBJECTIVES

Formulating the task of modification of an existing glass formula or developing a new one from the scratch one should consider accomplishing multi-task, which includes the following items:

- To find a formula superior to the existing one with combination of physical properties satisfying product-specific requirements.
- To provide glass "manufacturability", i.e. low-energy melting, acceptable forming range, low enough liquidus point, and others. There could be additional requirements specific for a given process, such as the melt density and surface tension, important for fiber and thin-sheet glasses.
- To ensure raw materials satisfy cost, keeping quality, and environmental safety requirements.
- To formulate free-to-operate or patent-able glass compositions.
- To know how minor variations of the glass composition affect manufacturing, and the final product properties.

METHODOLOGY

It was mentioned above that the problem of finding new glass compositions breaks into three categories: predicting properties by composition the most reliable way, multi-parameter optimization of physical properties and cost, and melting and characterization of all new glasses. *Simple and reliable method of calculation of properties by composition.* The method should enable glass scientists at the time of search, and the technologists at the time of trial and regular production to determine changes of critical technological and consumer properties of the material caused by even small compositional variations.

Reliable methods can help to outline areas of compositions containing, potentially, target glasses. The most efficient application of calculation methods, however, is to use them as a set of working functions (fitness functions and constrains) in an optimization program. This represents one of the main tasks of the present work.

There are known methods for calculation of glass properties by composition included in SciGlass[®]. (We should note that information system SciGlass[®] has been used as one of the major sources of glass properties data at the present study.) These calculation methods utilize, mainly, additive scheme considering contributions of the glass components in a given physical properties (density, viscosity, CTE, electric conductivity, refractive index, and some others) in form of members of polynomial equations of the second range. Disadvantages of the additive scheme are as follows. One – is a very high sensitivity of polynomial equations to normalization of chemical composition, so a minor error leads to gross errors. Two – that in order to add or replace glass components one need to re-build the equations. Three – the additive scheme represents interpolation, which uses as many available points as possible, which means low or no ability to sort the data points by their reliability.

There are several methods based on a "structural" approach. They consider glass or melt properties connected with structural groups or elements. The structure-properties relations in multi-component glasses, however, are quite complex and often unknown. This leads to free interpretation of properties changes, and higher uncertainty in their prediction. Known based-on-structure methods work well in case of glasses with a few components, and within a narrower compositional range.

We suggest a combined approach. The work begins with analysis of available data on a selected physical property within the range of composition, in which we assume certain types of structural groups determine this property (forming viscosity temperature T_3, for example). We begin with the widest possible ranges of components variation, and narrow them till most of the experimental points begin to follow the trend prescribed by a calculated line. Then we start iteration process aiming to find composition of the structural groups more accurately. Validation of the structural group(s) occurs by plotting together all available literature data and a linear regression with complex coefficient built using the sum of contributions of all structural groups in a value of the property of interest. One needs to reach the state, at which most of the literature data are getting close enough to the regression - within the error of measurements by using an analytical instrument of the second class. As we deal with a composition-property correlation we should take into account all random and systematic errors connected with determination of a physical property. Thus, deviation of data from their most probable values is connected with both the error of measurements and with errors of chemical analysis. In many cases glass formulas by batch are given, so exact compositions are unknown. The method of linear regression with complex coefficient is illustrated by figures below in the text.

The correlation gives us the ability to estimate a given property for any glass composition from the range with accuracy based on accuracy of available data, and the number of experimental points. We can exclude from consideration the data, located far away from the regression.

Multidimensional nonlinear constrained optimization of all key parameters: (cost, melting and forming temperatures, density, dielectric parameters, mechanical properties, $\Delta T_{3\text{-liq}}$, CTE, and others). Attempts to develop algorithms of global optimization have been taken many times. Some attempts were quite successful, but there is always a trend between probability of finding global minimums or maximums and number of iterations required for that. Computation time can be enormous, which is acceptable in many areas, but not in glass technology. We have tried to use several known methods of optimization of glass compositions by several independent parameters (physical properties, and cost), and, finally decided to develop a new one, based on state-of-the art mathematical routines (see below). The software program combining the new optimization algorithm with original methods of properties prediction allowed us to optimize fiber, sealing, and some other glass formulas inside and, even, outside the studied compositional areas. It is a possibility to do additional experimental search to verify or compliment the results of optimization.

Melting and testing begins with screening within the widest compositional range for finding sub-ranges with expected combination(s) of physical properties. The more detailed search within these sub-ranges allows identifying target ranges. The most accurate testing within these ranges helps finding prospective glasses-candidates, from which one to three can be used for commercial introduction.

REALIZATION

Property-composition correlations. Work begins with selection of glass system(s), which contain or may contain glasses having target characteristics.

We have selected CMAS (CaO-MgO-Al_2O_3-SiO_2) glass system, as it contains high-performance low-cost glasses, relatively large properties database, and no all the sub-areas studied in detail. We have outlined the following compositional area (in wt. %) to do our search: $3 < CaO < 30$; $2 < MgO < 25$; $1 < Al_2O_3 < 35$; $40 < SiO_2 < 70$.

Then one gets together all available literature data on the glass properties, and analyzes them in order to: outline correlations between key physical properties and composition varying within wide range, and to select the most reliable data to design reliable correlations for the further use. Glass and melt density, melting and forming viscosity, liquidus-to-forming window, tensile modulus and tensile strength are considered the most important characteristics of fiber glasses, for which correlations should be found. Once designed, the formulas are used to find glass compositions satisfying the following simple requirements: melting range viscosity $T_2 \leq 1450\,°C$, forming range viscosity $T_3 \leq 1280\,°C$, tensile modulus $E_t \geq 95$ GPa, tensile strength $\Box \geq 3.8$ GPa; density $d \leq 2.7 g/cm^3$, specific modulus $E_{sp} \geq 35$ GPa·cm^3/g, and cost of batch $c \leq 230$ units. Glass forming ability of the melts (to be overcooled to all-amorphous material) G_f, and ΔT_{3-liq} characterizing ability to stay crystal-free below T_3 during infinite time often can't be predicted from the literature data. Thus, we should begin our experimental work with validation of G_f for a large series of potentially high-performance glass formulas. Testing T_{liq}, and determination of ΔT_{3-liq} was added later to design corresponding regression, and increase chances to selects glasses-candidates with good manufacturability.

Below we consider building correlation for forming range viscosity (T_3). Modeling of T_3 is based on the following assumption. This property is determined by partial viscosities of structural groups formed by oxides glass-formers (acidic or amphoteric oxides), and modifiers (basic or amphoteric oxides). There are binary (silicates, for example), and mixed (alumo-silicates, for example) groups. Each group is presented by its structural formula, partial forming range temperature T_{3i} depending on energy of dissociation of the group, and weight factor k_{3i} determining contribution of the group in a property of the melt. Below, expression for complex coefficient of forming range temperature (K_{T3}) as a sum of contributions of several types of binary and mixed oxide fragments, and linear regression of T_3 with complex coefficient K_{T3} are given.

$$K_{T3} = \sum_{i=1}^{j} k_{3i} \times T_{3i} \times [M_mO_n] \times [A_oO_p]$$

$$T_3 = a + b \times K_{T3}$$

Where:

j – is a number of structural groups, which depends on a number of oxides net-formers and modifiers can vary from 2 to 9;

M_mO_n – is oxide-modifier, such as Na_2O, CaO, BaO, TiO_2, and others;

A_oO_p – is oxide-net-former, such as SiO_2, GeO_2, Al_2O_3, TiO_2, and others.

a, b – are coefficients of linear regression.

Fig.1 shows linear regression with complex coefficient K_{T3} obtained using literature data on forming range temperature T_3 measured by different authors for 126 experimental glasses of CMAS and MAS systems within compositional ranges of interest. Finding structural formulas and partial coefficients usually takes numerous iterations. Resulting correlation, very often, allows to get ~70% of available experimental points to follow a linear regression. Thus, one can see that only 33 points of 126 are outside of the $T_3(K_{T3})$ line. Remaining 93 points plus 12 points obtained by us in the course of preliminary study stay within the limit of error connected with own error of measurements plus error related to uncertainty of determination of glass composition by chemical analysis. Here we should note that in many cases the literature data were not accompanied by chemical analysis that increases the error.

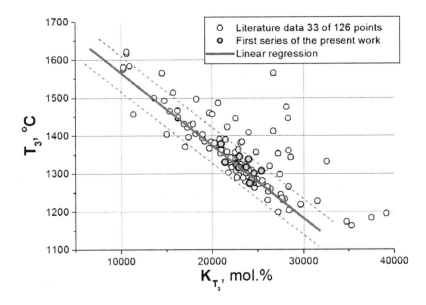

Figure 1: Linear correlation between forming range viscosity (T_3), and complex structural coefficient K_{T3} built using available literature data.

In order to build correlation between tensile modulus (E_t) and the glass composition we use slightly different structural interpretation of this property. E_t is a property of solid material, which depends on coordination and rigidity of the glass net, so the structural groups responsible for tensile modulus should be described in terms of coordination of atoms of net-formers, and the energy of chemical bonds.

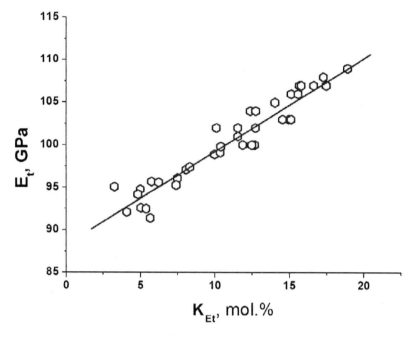

Figure 2: Linear correlation between tensile modulus (E_t), and complex structural coefficient K_{Et} built using available literature data.

These groups have been identified that allows us to build a proper correlation. The initial set of experimental data on E_t from several reliable literature sources measured by different testing methods is plotted on Fig. 2 alone with linear regression $E_t(K_{Et})$. In the course of the study the formula was validated by adding new sets of real experimental data measured with accuracy increasing due to larger number of test specimens to measure (from a single one up to 12), and more accurate testing method to use (ultrasonic method vs. sonic one).

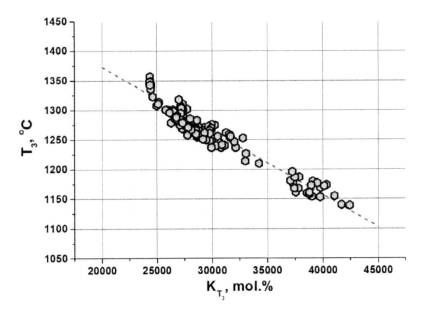

Figure 3: Actual data on forming range viscosity (T₃) measured for experimental glasses of this work plotted alone with updated linear regression obtained for this property as explained in the text.

Figs 3 and 4 show experimental data on forming range viscosity point (T_3), and tensile modulus (E_t) measured in the course of the present study. Although the points demonstrate certain deviation from the linear regressions (connected with errors of measurements, errors of chemical analysis, and errors of building the regressions), they allow finding the most probable values of these and several other physical properties (d, σ_t, T_2, ΔT, and others) for any glass composition within the range of interest, and make it easier to design new series of experimental glasses.

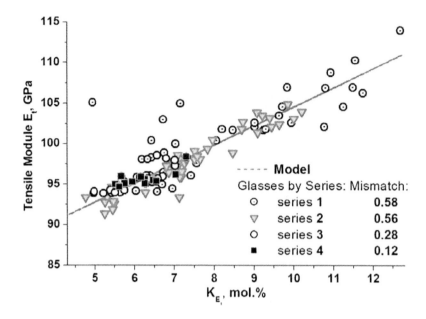

Figure 4: Actual data on tensile modulus (E_t) measured in this work for over 300 of experimental glasses of the first, second and third series by sonic (contact), and fourth series by ultrasonic (non-contact) methods plotted alone with a linear regression obtained for this property as explained in the text.

Fig. 5 shows correlation between $\Delta T_{3-liq} = T_3 - T_{liq}$ and its complex coefficient. Values of T_{liq} in this work were measured by express testing method, which included long-time discrete high-temperature holds of thin-layer specimens followed by quick cooling to room temperature followed by microscopic observation. We should note that T_{liq} is a property of molten glass, which can be determined with error connected with many factors (types of crystalline phases and crystal growth / melting rates, size and geometry of a boat or cell, nucleation regime, crystallization regime, temperature measurements, etc.), so one shouldn't expect a very high precision of determination of ΔT_{3-liq}. Such correlation, however, allows choosing glasses with higher positive values of ΔT_{3-liq} (lower crystallization ability) as potential candidates for the further search.

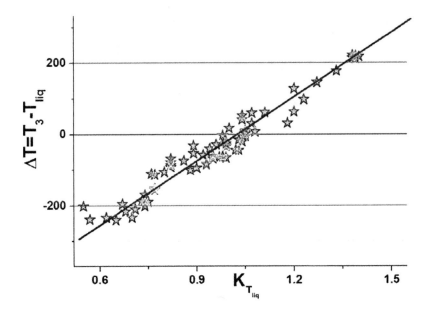

Figure 5: Linear regression with complex coefficient obtained for forming range "window" $\Delta T_{3-liq} = T_3 - T_{liq}$ obtained for a large series of the molten glasses studied in this work. In spite of significant scatter of the experimental points, one can sort all the glasses out into categories from "non-workable" (large negative numbers) to "excellent" (large positive numbers) determining the glass manufacturability.

As the correlations between composition and several important glass (melt) properties become available one can begin choosing compositional sub-ranges, potentially, meeting certain search criteria. Although, the correlations between properties, and composition are developed to be a part of optimization process (see below in the text), one can begin design of trial glass compositions with a simple procedure of outlining areas on ternary diagrams where key parameters have proper levels, and finding zones of their overlapping. Fig. 6 contains, as an example, a set of ternary diagrams with areas occupied by glasses having glass-forming ability (a), batch cost below 230 units (b), forming range temperature below 1280°C (c), and specific tensile modulus above 35 GPa*g/cm³.

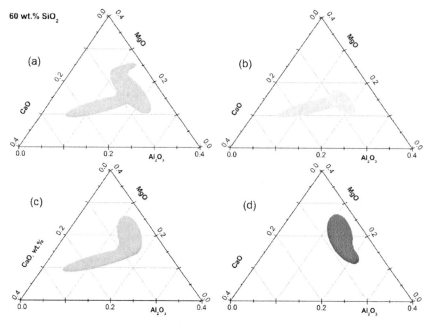

Figure 6: Compositional areas in a cut of CMAS system corresponding to silica content 60 wt. % built as a result of properties calculation (except for glass forming ability G_f) satisfying the following criteria: glass-forming ability G_f positive (a), batch cost $c < 230$ units (b), forming range temperature $T_3 < 1280°C$ (c), and specific tensile module $E_{sp} > 35$ GPa*cm³/g.

As it was mentioned above, the ultimate task for property-composition correlations was to include them in optimization program as sets of functions (fitness function – constrains) used, along with batch cost, for formulating glass formulas with optimal combination(s) of properties (see below).

OPTIMIZATION METHOD

Mathematical optimization is often used in various fields of applied science. Despite the significant progress made over a century of research, there is still no universal way to treat all variety of problems one can encounter. One faces a typical problem of finding global extremum of function: $\min_{x \in G} f(x)$, in a specified region of parameter space $G \subset R^n$. Particularly, function $f(x)$ (commonly called fitness function) represents a specific property of glass (tensile modulus, density, melting and forming range temperatures, forming-liquidus window, cost of production, etc.). Whereas x determines composition of the glass: $x = (n_{Al_2O_3}, n_{CaO}, n_{TiO}, n_{SiO_2}, n_{BaO} \ldots)$ is a N-component vector, where N equal or more than 4. Conditions $0 \le n_i \le 1$, and $\sum_{i=1,4} n_i = 1$, restricts x to lie within N-dimensional tetrahedron T_N.

One, usually, optimizes a given property constraining others. For example, it is very natural to put constraints on the cost of glass. Thus, generally the region of search G is compact and lies within mentioned tetrahedron.

Depending on the fitness function, different types of optimization algorithms should be used. Usually, the best performance is obtained by *"interior point method"* algorithms. One needs, however, to make sure that fitness function is smooth enough or continuous with at least it's first derivative. In our case we encounter two types of functions. The first type represents analytical expressions (usually rational functions). The second type of functions represents result of optimization themselves. For example: liquidus temperature at the moment is given by a neural-network simulated approximation of experimental data. Cost of the glass is a solution of linear programming problem. Both batch cost, and liquidus temperature are very important. Thus, in general we have to consider "black-box" type functions.

Another important point is that there is always a finite error in the composition of manufactured glass. Therefore, deviation from the target composition is supposed to lie within the range of 1 percent for each oxide type. This somehow simplifies the problem.

Taking into account all assumptions mentioned above, we've developed an algorithm, which was proven to give global extremums. First, the whole domain is divided into smaller tetrahedrons dT_i with 0.01 edges in each dimension. Within of each dT_i, such that $dT_i \cap G \ne 0$, the conjugate-gradients descents or direct searches are performed, depending on the types of functions used. Then, the resulting local extremums are compared to give the global extremum value. Though time-consuming, this approach allows dealing with not simply-connected search domains G, but gives a physically-justified global extremum.

Estimation of cost of glass composition is done using "revised simplex method". Obviously, linearity of the problem itself along with the algorithm used ensures that the provided solution will correspond to global extrema. As a remark, we provide here the formulation of this sub-problem. In this case, fitness function $p = \sum_{k=1,N} b_k c_k$ is a sum of all available batch components, where b_k is a mass proportion of k-th component in resulting batch, where c_k is it's price per ton. The constraints are:

$$n_i - 0.005 \le \sum_{k=1,N} b_k x_{ik} \le n_i + 0.005, i = 1..3 ; \quad \sum_{k=1,N} b_k = 1,$$

Where: matrix element x_{ik} represents the portion of oxide i in raw component k. The error deviation is introduced to ensure feasibility. Surprisingly, we have found that implementation of this textbook algorithm in different commercially available packages doesn't give the same result, whereas there is no freedom for such deviation.

In order to make the process of designing new glass compositions more efficient we have developed a software program called "Opimax 1.0". The "Opimax 1.0" is an utility program designed to perform optimization tasks and data visualization oriented on 4 to 8 component glass systems. Specifically, the main tasks can be formulated as follows:

1. Model input (model is a set of user defined functions depending on glass composition);

2. Ability to plot functions defined in the model within the whole compositional area (tetrahedron);

3. Constrained function (properties and batch cost) optimization with visual representation of the results on multi-graph screen(s);

4. Glass price calculator built-in utility.

The program has a text-based input, which means that the main task will be formulated in the form of a "script" or a "program". To define model and perform operations one has to know the scripting language syntax. It is fairly simple and can be summarized in a set of simple rules, so the user can easily adjust such important elements as coefficients in property-composition correlations, constrains, and boundaries (concentrations of glass components).

At this time optimization software doesn't include correlations between glass composition and its glass forming ability. The reason for that is the absence of simple a reliable structural approach for G_f, which, from one hand, leads to uncertainty in placing inter-phase boundaries, and, from another hand, probability to skip some potential glasses-candidates as crystallized materials. Thus, the results of optimization should always be accompanied by G_f test.

The program contains pre-defined expressions for glass properties and batch (glass) cost, (specific for each new composition exploratory project). The users can modify the existing functions or develop their own formulas obtained as a result of totally new search or post-development work.

EXPERIMENTAL WORK

Melting and testing work begins with determination of glass forming ability (G_t). Although studies of the glass forming ability in a number of multi-component systems were performed and the results (areas on diagrams) published, one can find prospective areas, for which real experimental points are little or absent. Thus, repetition of glass forming ability test with accurate batching using set of specific raw materials is quite useful thing to do. Examples of custom ternary phase diagrams with the areas containing glass-forming melts in quaternary CMAS system obtained in the course of development of method for identifying of high performance glasses are presented. One can see that the ranges found in present work don't necessarily coincide with those found in the literature.

It can be explained by the fact that phase diagrams were created by multi-disciplinary groups using different techniques, raw materials, and with errors of determination of phase borders. The experimental work on glass forming ability has been performed by melting small amounts (200-300 mg) of glasses from pure oxides at 1500°C and cooling them to obtain samples for observation under precision microscope. A sample considered amorphous if not a single crystal was registered under 400^X magnification. Crystallization ability quickly estimated by this test was validated by more accurate methods as the program progressed further

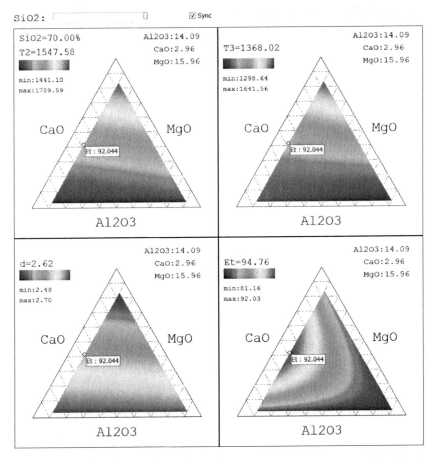

Figure 7: Output window of graphic user interface of Optimax 1.0 program. The following results of computation and optimization are available: 1). Determination of four properties at the same time (T_2, T_3, E_t, d); 2). Compositional range (in wt. %) is set as follows: $3 < CaO < 30$, $2 < MgO < 25$, $1 < Al_2O_3 < 25$; $40 < SiO_2 < 70$; 3). Optimized parameter – E_t to find the global (within the whole 4D area), and local (for each level of SiO_2 – can be changed by moving slider in the upper left part of the screen) maximums of E_t. $E_t = 92.044$ GPa represents a local maximum for $[SiO_2] = 70$ wt.%. Note: the calculations were made by using demo correlation functions.

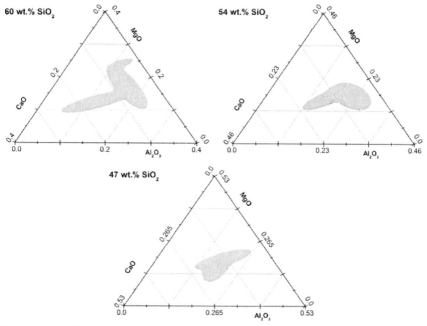

Fig. 8. Compositional areas corresponding to glass-forming melts on ternary phase diagrams representing sections of tetrahedron of quaternary CMAS system at 47, 54, and 63 wt. % of SiO$_2$.

As a result we create the first matrix containing numerous glasses to choose those passing glass-forming ability test. Physical properties (T$_2$, T$_3$, E$_t$, d, σ_t) and batch cost of these glasses should be estimated using property-composition correlations (see above). By analyzing areas on ternary diagrams one can select candidates for the further experiments. One takes larger portions of batch materials for melting and express protocols for testing. Thus, the glass viscosity is tested by Archimedes (counterbalance) method using small platinum sphere and small crucible, tensile modulus is measured by using sonic method, density is determined by hydrostatic weighting, liquidus temperature is determined by step-wise crystallization in a thin layer, tensile strength is measured by ring-in-ring test. Each test is performed using a single specimen per glass composition, that allows testing several glasses a day. Results of the work include: database with properties of dozens of glasses, and new or updated property-composition correlations. Next series of glass compositions is composed by using the optimization software. This series includes as many glasses as needed to select a few candidates for the final validation. Melting experimental glasses, and testing their physical properties is carried out using more precise methods or/and instruments. Thus, to improve the accuracy of measurements of E$_t$ one should perform contact sonic test on up to 12 individual test specimens or use the more precise non-contact ultrasonic method. Having additional physical properties data one can make the final selection of three glasses-candidates, which should be melted to ensure uniform, on-target, low-defect samples suitable for making test specimens for standard extra-precision testing. We should note that depending on the task additional iterations may be needed to reach the target combination of cost and properties.

SUMMARY

1. Although numerous studies on development of new glass compositions have been performed, and many papers and patents published there are still "white spots", which, potentially, contain high-performance glass formulas;

2. In order to identify these spots one should start the work from-the-scratch skipping no steps in searching for new glasses;

3. Combining experimental work with innovative exploratory tools, such as multidimensional global optimization program containing advanced property-composition correlations, increases efficiency of compositional search;

4. It could be useful to investigate wider compositional areas, as well as uncommon combinations of oxides glass-formers and modifiers to break-through in compositional search;

5. Although surface modification by ion-exchange treatment or coatings provide many benefits, there is still a room for improvement by changing chemical compositions of glass;

6. Joining effort of researchers and technologists is often produce synergetic effect, so the target glasses are developed faster.

Melting, Raw Materials, Batch Reactions, and Recycling

ADVANCED MELTING TECHNOLOGIES WITH SUBMERGED COMBUSTION

I.L. Pioro[1], L.S. Pioro[2], D. Rue[3], and M. Khinkis[3]

[1] Faculty of Energy Systems and Nuclear Science, University of Ontario Institute of Technology
2000 Simcoe St. N., Oshawa, Ontario, L1K 7H4, Canada
[2] Advanced Energy Technologies Consulting (AETC) Inc., Oshawa, ON, L1L 0A4, Canada
[3] Gas Technology Institute, 1700 South Mount Prospect Rd., Des Plaines, IL 60018-1804, USA

ABSTRACT
This paper is devoted to the development and design of advanced melting technologies with submerged combustion. The objective is to compile and summarize findings and experiences of scientists and engineers from various research organizations and companies in this new area worldwide. The main advantages of submerged combustion, i.e., the combustion of gas-air or gas-air-oxygen mixtures directly inside a melt, are to achieve the maximum heat transfer from combustion products to the melt, to improve mixing, i.e., melt homogeneity, and to increase the rate of chemical reactions. Based on these advantages of submerged combustion, advanced melting technologies were developed and tested in various industries. The paper overviews the development, design, testing and industrial application of various advanced melting technologies with Submerged Combustion Melters (SCMs) for melting silicate materials; vitrification of high-level radioactive wastes and some other applications.

1. INTRODUCTION

Submerged combustion [1], as its name implies, is the combustion of gas or fuel oil in such a manner that the hot combustion-product gases are released under the surface of liquid or melt. In this way, the energy released by the combustion process is transferred by direct contact with the liquid or melt. Therefore, two major classes of submerged combustion devices/apparatuses can be identified:

1. Submerged combustion devices (SCDs) for a liquid heating and/or evaporation.

    In such devices, gas- or fuel-oil-air mixture used for the combustion process. Most SCDs are arranged with a burner above the liquid level and a submerged exhaust system. Although it is possible for the burner itself to be submerged into the liquid [2]. Submerged combustion is used in two classes of evaporators direct and indirect. In the first, it is used to concentrate corrosive or toxic materials. In the second, water is heated, which in turn, is then circulated over a tube bank containing the liquid to be evaporated. Advantages of submerged combustion are: (a) achieving maximum heat transfer rate from combustion products to the liquid; (b) the absence of fouling or corrosion; and (c) the ability to handle highly viscous liquids or liquids containing up to 40% solids. However, a disadvantage is a contamination of a liquid with combustion products.

2. Submerged Combustion Melters (SCMs) [2] for a solid or powder-like charge heating and melting.

    In these apparatuses, gas-air mixture or gas–air-oxygen mixture used for the combustion process. SCMs are always arranged with a burner submerged into or beneath the melt. The main goals of submerged combustion in this application are: (a) achieving maximum heat-transfer rates from combustion products to the melt; (b) improving mixing (i.e., melt homogeneity); and (c) increasing a rate of chemical reactions inside SCMs. Currently, various SCM designs have been developed for producing materials for the building industry from metallurgical slag [3], coal slag and ash from coal-fired thermal power plants [4, 5]; fuming of slags of non-ferrous metals [2]; melting silicate materials [2]; producing mineral wool [2]; producing molten defluorinated phosphates for agriculture [6]; pyrohydrolysis of fluorine-containing wastes [2]; vitrification of high-level radioactive wastes [7-10]; and production of expanded-clay aggregate for lightweight concrete from non-selfbloating clays [11]. Many of the developed technologies are intended for decreasing harmful effects of various wastes such as slags, ash, etc. on the environment by effectively reprocessing them into materials for the building industry or by safe infinite disposal of high-level radioactive wastes by including them into a glass matrix [2].

In the submerged-combustion process (see Fig. 1) the fuel (natural gas) and oxidant (air or air-oxygen mixture) are fired directly into the bath of materials being melted. The high-temperature bubbling combustion inside the melt creates a complex gas-liquid structure and a large heat-transfer surface. This significantly intensifies the heat transfer between the combustion products and the processed material. The intense mixing of the melt increases the speed of melting as well as the speed of chemical reactions of mineral formation, and improves the homogeneity of the molten product. Another positive feature of the SCM is its ability to handle relatively non-homogeneous charging materials. The granular structure and, especially, the homogeneity of the charge do not require strict control. Charge components can be loaded either premixed, separated, continuously, or in batches.

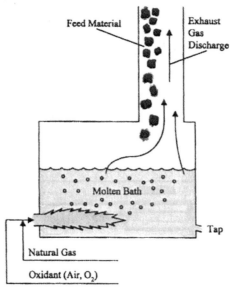

Figure 1. SCM concept.

One of the main conditions that must be satisfied for a successful operation of the SCM is the achievement of stable combustion inside the melt. It might seem sufficient for stable combustion to supply a combustible mixture of fuel and oxidant into the melt, the temperature of which significantly exceeds the ignition temperature of the fuel. However, numerous experiments conducted in various submerged combustion furnaces with different melts have shown that simple injection of a combustible mixture into a melt (for example, through lances) does not ensure stable combustion. Instead, it results in formation of cold channels, which lead to explosive combustion and excessive melt fluidization.

Results show that for the majority of melt conditions the ignition of a combustible mixture injected into the melt starts at a significant distance from the injection point. This leads to the formation of cold channels of frozen melt and explosive combustion. To avoid this type of combustion process, the system must be designed to minimize the ignition distance. This can be achieved in three ways: (a) by stabilizing the flame at the point of injection using special stabilizing devices, (b) by splitting the fuel-oxidant mixture into smaller jets, and/or (c) by preheating the fuel-oxidant mixture.

Based on the characteristics of submerged combustion, the burners must be designed to provide:

- Stable flame at the point of injection of the fuel-oxidant mixture into the melt to prevent the formation of frozen melt and resultant explosive combustion.
- Constant, reliable, and rapid ignition of the fuel-oxidant mixture so the mixture can burn quickly and release the heat of combustion into the melt.
- Highly uniform distribution of the fuel-oxidant mixture across a cross section of the melt bath to ensure maximum gas-liquid heat-transfer surface.
- Long service life, capability for a wide range of turndown to match the thermal load, no flashback potential, and simple maintenance.

The Gas Institute of the National Academy of Sciences of Ukraine (NASU) (Kiev, Ukraine) has designed several types of multiple-nozzle burners that meet these requirements. Figure 2 shows one of these designs. The burner uses a welded metal casing and a partitioned, water-cooled combustion chamber. The nozzles connecting the combustion chamber to the combustion-air plenum are located on the side walls of the combustion chamber. These nozzles have also additional openings connected to the natural-gas plenum. The mixing of natural gas and combustion air initiates within the nozzles and is completed in the combustion chamber. The flame is stabilized in the recirculation zone within the combustion chamber because of intense mixing of the combustion products with fresh portions of the fuel-oxidant mixture. From the combustion chamber, the stable burning fuel-oxidant mixture is injected into the melt, and combustion is completed in bubbles formed inside the melt above the burner. The energy in the flame jets and the rising bubbles provide intense mixing and homogenization of the melt. The burner is attached to the bottom of the bath so its main body is outside the furnace. Only the surface around the exhaust of the slotted combustion chamber is in contact with the melt.

2. REPROCESSING METALLURGICAL SLAG INTO MATERIALS FOR BUILDING INDUSTRY INSIDE SCM

Several methods of reprocessing metallurgical (blast-furnace) slag into materials for the building industry, based on melting apparatuses with submerged combustion, were developed and tested. The first method involves melting hot slag with some additives directly in a slag ladle with a submerged gas-air burner with the objective of producing stabilized slag or glass-ceramic. The second method involves direct draining of molten slag from a ladle into a slag receiver with subsequent control of the slag draining into the SCM, where special charging materials are added to the melt with the objective of producing glass-ceramic. A third method involves melting cold slag with some additives inside an SCM with submerged gas-air burners with the objective of producing glass-ceramic fillers for use in road construction.

Specific to the melting process is the use of a gas-air mixture with direct combustion inside the melt.

The reprocessed blast-furnace slag in the form of granules can be used as fillers for concretes, asphalts, and as additives in the production of cement, bricks and other building materials. As well, reprocessed blast-furnace slag can be poured into forms for the production of glass-ceramic tiles.

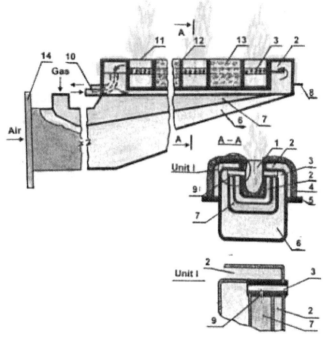

Figure 2. Bottom submerged combustion burner: 1 – combustion-stabilization chamber, 2 – water jacket, 3 – air nozzles, 4 – lining, 5 – flange, 6 – air duct, 7 – gas duct, 8 – flange, 9 – gas nozzles, 10 – water in and out, 11 – combustion chamber casing, 12 – partition, 13 – lining, and 14 – air-duct flange.

2.1   Introduction

Much slag is created during operation of metallurgical plants around the world. In general, these solid wastes are partially reprocessed; however, a significant amount of them remains, damaging the environment. Reasons for under-usage of metallurgical slag are well known, including difficulties in reprocessing due to inconsistencies in slag chemical content; a relatively high cost of transportation, and a high cost of slag re-melting.

One of the ways that allows the most complete reprocessing of metallurgical slag is the development of SCMs for reprocessing slag into effective building materials in a special technological process and location of these SCMs near a metallurgical furnace.

In general, slow-cooled (free-convection air cooling) metallurgical slag undergoes crystallization and, after time, disintegration due to its non-uniform content. However, if the metallurgical slag undergoes fast cooling, such as cooling with water (direct draining into a water pool or water-spray cooling – so-called "water granulation"), its structure becomes more amorphous (glassy slag).

The fast water-spray cooling of the molten slag creates small, hard glasslike granules. These granules have a very wide range of application as fillers for concretes and as additives in the production of cements and bricks. Melting metallurgical slag in melting apparatuses improves its uniformity and quality, widening its application in the building industry.

2.2 Innovative Technologies for Reprocessing Metallurgical Slag into Building Materials

2.2.1 Reprocessing metallurgical slag inside slag ladle with submerged burner

It was proposed to reprocess hot metallurgical slag directly in a slag ladle. The main ideas of this new engineering solution were: 1) to submerge a gas-air burner inside the molten slag and 2) to add into a slag ladle a required amount of quartz sand with some additives to obtain a melt with predetermined physico-chemical properties. After that, the melt should be drained into a glass furnace/melt receiver for subsequent continuous pouring into forms.

The experimental set-up (see Fig. 3) consisted of an internally lined firebrick canopy (5) and a movable water-cooled burner (3). A combustion gas-air mixture (coefficient of excess air 0.9–1.1) was supplied through an internal tube. A cart (1) with a slag ladle (4) was pulled under the canopy (5) and the gap between the ladle (4) and canopy (5) was covered with firebricks. A burner (3) was submerged into the molten slag (2) with the burner penetrating through the thin crust. Combustion products bubbled through the melt (2) and were removed via an exhaust pipe (7). Depending on the desired physico-chemical properties of the final product, the corresponding charging materials were loaded through an opening (9) in the canopy (5).

To obtain only stabilized slag (chemical stabilization required), quartz sand was added to the molten slag and the melting process was continued for about 45 minutes. To obtain the melt for production of glass-ceramic, various additives were used in quantities from 900 to 2500 kg per ladle (or 9 – 25% by weight). After loading the additives, the melting process took up to 45 minutes. The final product was glass with predetermined physico-chemical properties uniformly distributed throughout the entire melted volume. Even assimilation of barely dissoluble additives was within the predetermined ranges. However, an excessive melting loss was noticed for fluorine and sulfur.

Glass samples obtained during ladle-slag melting with about 50 kg/t of the charging materials (sand, clay, sodium sulphate, sodium thiosulphate, etc.) were subjected to thermal treatment in a laboratory. Analysis of these glass-ceramic samples showed good crystallization and homogeneity.

2.2.2 Design of experimental metallurgical slag reprocessing line

An experimental metallurgical-slag reprocessing line was designed by the Gas Institute NASU, and built at the Avtosteklo glass plant (Konstantinovka, Ukraine) with the objective of investigating various aspects of the technological process of slag reprocessing inside an SCM with submerged combustion. The experimental slag-reprocessing line consisted of a slag-melting furnace (A), an SCM (B) and a melt-clarification furnace (C) (see Fig. 4).

The reprocessing line operated as follows (Fig. 4): cold slag (10) was loaded into the slag-melting furnace (A) and melted there. The molten slag then drained through the slag-chute (2) into the SCM (B), where some additives were loaded through window (8). To decrease entrainment of additives, the charging materials were slightly wetted when added into the melt. The melt was drained into the melt-clarification furnace (C) from which it was poured into forms to produce glass-ceramic tiles.

Figure 3. Reprocessing of molten slag directly in ladle: 1 – slag cart, 2 – melt, 3 – submerged burner, 4 – ladle, 5 – canopy, 6 – third servicing level, 7 – exhaust pipe, 8 – gas, air and cooling-water inlets, and 9 – charging-materials loading hatch.

The line was designed for continuous operation with a production capacity of 400 kg/h. The melts produced in the SCM corresponded to the predetermined chemical content. To increase the overall efficiency the combustion products from the SCM were directed through the slag chute into the slag-melting furnace. This feature decreased gas consumption for the slag-melting furnace by 10 – 15%. Successful operation of this experimental reprocessing line proved its high efficiency in reprocessing cold metallurgical slag into glass-ceramic tiles.

An innovative SCM design (Fig. 5) for reprocessing slag at a metallurgical plant (Mariupol', Ukraine) was developed by the Gas Institute NASU together with several design institutes. The SCM was intended for continuous reprocessing of hot metallurgical slag by the above-mentioned technology. The operational capacity was up to 250 tons per day.

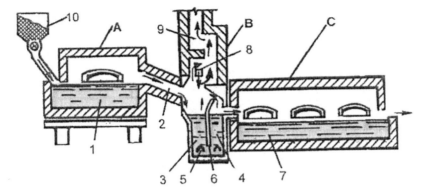

Figure 4. Schematic of experimental metallurgical slag reprocessing line: (A) slag-melting furnace, (B) SCM, and (C) melt-clarification furnace; 1 – molten slag, 2 – slag chute, 3 – primary melting chamber, 4 – secondary melting chamber, 5 – submerged gas-air burners, 6 – partition, 7 – molten glass-ceramic, 8 – loading window, 9 – flue-gases duct, and 10 – crashed cold slag.

Figure 5 shows a basic design of the SCM. One of the specifics of the SCM is the utilization of heat from the converter flue gases to heat the hot slag inside the receiver (1). The receiver (1) was fixed on two trunnions (9) carried on rollers (2). Molten slag from a ladle is poured into the receiver through an interim funnel (10) into an opening in the upper part of the receiver. The amount of poured slag into SCM (3) is proportional to the inclination angle of the receiver, making it easy to control the slag loading process. The amount of charging materials loaded inside the SCM is regulated with a weighing-hopper. Charging materials are loaded inside the SCM with the feeder through opening (8) under the melt level, thus significantly decreasing entrainment of the dusty portion of the charging materials. The SCM is heated with gas-oxygen lances (4). The prepared melt is drained through a tap-hole (5).

To increase the overall thermal efficiency a waste-heat thermosyphon boiler [12] for a hot-water supply (Fig. 6) can be installed in a flue-gases duct downstream of the SCM.

2.2.3 Reprocessing metallurgical slag into road glass-ceramics
Road glass-ceramic – "Dorsil" – developed at the State Institute of Glass (Moscow, Russia) is a building material intended to be used as a filler of asphaltic-concretes and acid-durable concretes. Blast slag and sand are used as the raw materials to produce these glass-ceramic fillers. For glass-ceramic-filler production, a glass mass uniformly saturated with gas inclusions is required. After thermal treatment in a crystallizer, this mass acquires the required strength and whiteness.

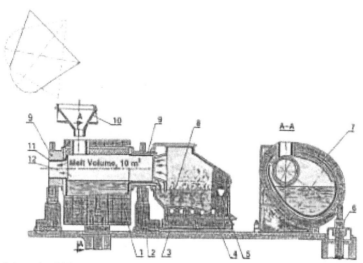

Figure 5. Schematic of SCM aggregate for reprocessing molten slag and charge into building materials with capacity of 250 tons per day: 1 – slag receiver, 2 – supporting rollers, 3 – SCM, 4 – openings for lances, 5 – tap-hole, 6 – hydraulic jack, 7 – refractory, 8 – charge loading, 9 – receiver trunnions, 10 – interim funnel, 11 – melt lining, and 12 – flue gases.

(a)   (b)

Figure 6. Waste-heat thermosyphon boiler for hot-water supply: (a) single unit overall view, and (b) multiple units installed inside flue-gases duct.

Technical characteristics of developed glass-ceramic fillers follow:

Bulk weight	$1300 - 1400$ kg/m^3;
Loading weight	$1000 - 1100$ kg/m^3;
Water absorption	$3 - 5\%$;
Compression strength	$88 - 98$ MPa;
Frost resistance	more than 300 cycles.
Chemical durability:	
In H$_2$SO$_4$	not less than 99%;
In HCl	not less than 96%.
Integral coefficient of reflection	$52 - 55\%$.

An experimental glass-ceramic filler production line with capacity of 1000 tons per year was constructed and successfully operated at the Experimental Glass Making Plant (Tula, Russia). This production line included: an SCM (Fig. 7) designed by the Gas Institute NASU, a crusher and a rotating furnace for crystallization. Figure 8 shows a newer design of the two-phase thermosyphon vault of the SCM.

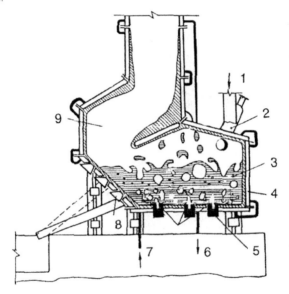

Figure 7. Schematic of SCM for glass-ceramic filler production: 1 – charge loading, 2 – feeder, 3 – melt, 4 – SCM body (water jackets), 5 – burners, 6 and 7 – cooling water inlet and outlet, 8 – tap-hole for melt draining, and 9 – flue-gases duct.

Selection of an SCM as the melting unit was based on:

- SCMs are characterized by high specific productivity, exceeding by one order of magnitude the productivity of bath glassmaking furnaces;
- SCMs are able to produce white, non-sulfurous glass-crystalline material with a porous, rough surface without additional foam-making additives; and
- SCMs do not require a special pre-treatment of the charging materials.

A homogeneous glass mass containing $25 - 60\%$ by volume of gas inclusions is produced in the SCM. The melting temperature reaches about $1600°C$. The specific productivity of about 15 tons of melt per 1 m^2 of melt free surface was reached for SCMs with a working volume of $4 - 8$ m^3 and a melt free surface area of $2 - 4$ m^2.

Figure 8. New design of SCM vault with two-phase thermosyphons for production of glass-ceramic fillers.

The glass-ceramic fillers produced were mainly used as an additive to asphaltic-concretes for constructing upper layers of clarified pavements. Asphaltic-concrete pavements with glass-ceramic fillers have such advantages:

- Clarified surface of the pavement improves esthetic image of city streets and corresponds to modern requirements for decorative design;
- Light-color pavement reflects a significant portion of solar energy, therefore improving the microclimate of streets. Also, the costs of electrical energy used for street and highway lighting in the evening and night time decrease; and
- Using asphaltic-concrete pavements with glass-ceramic fillers increases pavement roughness, therefore decreasing vehicle braking distance and improving traffic safety.

Glass-ceramic fillers are a relatively light material and usage of it as an additive instead of granite or other fillers in various concretes leads to their bulk weight decreasing. Glass-ceramic fillers are also successfully used in chemically resistant concretes and in polymeric concretes for protecting flue-gases ducts in coal-fired power plants from the destructive action of sulphuric acid. In addition, acid-resistant powder and facing-glass aggregate (crumb) of various colors can be obtained based on the porous glass-crystalline material. The glass aggregate (crumb) has high resistivity to atmospheric exposure, abrasion, impact, and to the destructive effects of water, acids, and alkalis. The wide palette of colors (white, light brown, green, yellow, blue, dark red etc.) also provides success in using glass aggregate for facing facades as well as for lining buildings.

The proposed SCM produced 980 tons of glass-ceramic fillers and 560 tons of glass grit in just one year. The latter product was used as the acid-resistant filler for building flue-gases ducts at coal-fired power plants.

3. SCM FOR MINERAL WOOL PRODUCTION

The SCM technology was proposed for production of mineral wool instead of cupolas. Two 75 t/d SCMs are currently in operation for mineral wool production: one in Ukraine and the other in Belarus'. These commercial melters use recuperators to preheat combustion air of up to 300°C. Both melters operate with less than 10% excess air and produce NO_x emissions of less than 100 vppm (corrected to 0% of O_2) along with very low CO emissions. Work is underway at the Gas Institute NASU to improve both the thermal efficiency and the specific production rate by increasing the combustion air temperature to 400°C and incorporating a heat-recovery charge preheater.

3.1 Introduction

The SCM is a bubbling bath-type furnace that shows promise for a number of processes to produce glass melts from various feed materials (geological rocks, sand, waste slag, ash, etc.). One of these applications is the production of mineral (silica) melts for the manufacture of thermo-insulation fibers. These mineral melts are produced in various types of thermal units. The most widespread are cupolas, while some reverberatory (tank) and electric furnaces also are used. Although cupolas are effective in producing acceptable melts, their operation requires expensive coke and a granulated charge. In addition, the combustion products contain high levels of carbon monoxide, hydrogen sulfide, and dust, and the melt produced is not very homogeneous.

The tank furnaces are relatively large and consequently require large installation footprints and periodic replacement of large amounts of high-quality refractories. These refractories also have relatively high heat capacity, requiring long-term continuous operation. Similar to cupolas, both tank and electric furnaces also require high-quality, specially prepared charges. Additionally, the electric furnaces operate at relatively low energy efficiency when the actual amount of fossil energy consumption is taken into account.

To address the above drawbacks of conventional technologies, the Gas Institute NASU has used the principle of submerged combustion to develop a compact converter-type melting furnace that provides high-production rate while using very little refractory.

3.2 Basic SCM design

The design and operation of SCMs require special care to minimize the fluidization of the melt, which creates a large number of droplets. These droplets, especially the small ones that are formed when the bubbles split, are thrown out of the melt to a significant height, sometimes over 10 m. Consequently, those who are engaged in the development of these submerged-type combustion melters must explore ways to protect the exhaust ducting from being covered by the frozen melt. Some designers have addressed this issue by increasing the height of furnace space above the melt to minimize the amount of melt reaching the exhaust duct. In the Gas Institute NASU design, this issue is resolved by reducing the height of the melt bath to the minimum acceptable levels and by removing the combustion products through a special separation zone. This approach reduces the necessary water-cooled surface area around the melting zone.

A key factor in designing the SCMs is an optimization of the geometrical dimensions and configurations of the melting/combustion zone. Based on research data, the Gas Institute NASU has developed a technique for specifying the dimensions of the melting bath, including the optimum depth of the bath, for uniform distribution of the combustion gases. This technique allows estimation of optimum dimensions to maximize the ratio of heat absorbed by the melt to heat transferred to the cooling water.

In the melting bath, the heat transfer between the high-temperature combustion products and the charge particles primarily occurs through the melt. For convenience, this process can be divided into two steps: (a) the heat transfer between the combustion products and the melt and (b) the heat transfer between the melt and the charge particles. Studies have shown that the first step is controlling when producing melt for the production of mineral wool. If the charge contains a significant quantity of quartz sand (>40%), however, the second step (dissolution of SiO_2) becomes controlling.

3.3   Commercial Operation

The industrial SCM design, illustrated in Fig. 9, is a result of extensive research efforts at the Gas Institute NASU, the highlights of which are discussed above. Its main components include melt bath, separation zone, recuperator, feeder, melt tap-hole, submerged burners, and stack. In addition, to ensure reliable and steady operation, it employs systems for tank cooling, natural gas and combustion air supply and process control, measurement, and safety.

The melt bath is assembled from separate carbon-steel water-cooled panels lined with a 30-mm layer of refractory on the surface exposed to the melt. The panels are divided into groups, and each group is connected to the cooling-water system, which includes a cooling tower. The submerged burners are connected to the bottom of the floor panel. The loading of material is done through a port in the crown panel by a loading and proportioning device. This device is designed to not only feed the charge in proportion to the production rate, but also to protect the feed opening from closing in the presence of the fluidized melt droplets.

The flue gases from the melting bath are passed through a separation zone, where the melt droplets and solid particles are separated as a result of the centrifugal forces that are created by the turning flue-gases flow. The flue gases go into the recuperator and then into the stack, while the droplets and the solid particles remain in the furnace.

The recuperator is designed to preheat the combustion air to a temperature of about 300°C. The recuperator is a basket type, and the majority of the heat transferred to the tubes from the flue gases is by radiation.

The melt is discharged from the melting bath using a special tapping device. Different designs are used depending on the operating mode of the melter: periodic or continuous. For mineral wool products, the tapping device is designed to ensure melt discharge in a continuous stable stream. The melt is then transported along an inclined heated channel to the fiber-forming device (centrifuge).

Currently, two SCMs based on the above design are operating at two separate plants in Kiev (Ukraine) and in Beryoza (Belarus'). Both produce thermo-insulation mats from mineral fiber. Both have an area of 4 × 5.5 m, and their height to the top of the recuperator is 9 m. The performance characteristics of the SCM in Kiev are listed under option A in Table 1. The production rate from this melter is about 500 – 550 m^3 of mats per day with a density of 100 kg/m^3. The waste is negligible since the 12 – 18% rejects are fed back into the melter. The melt contains small gas bubbles that have not been found to have any adverse impact on the quality of the mineral fiber.

The flue gases leaving the melter contain only negligible amounts of carbon monoxide and only 140 – 200 mg/m^3 of NO$_x$ (corrected to 0% O$_2$). The NO$_x$ levels are less than one-quarter of those generated in the best conventional tank melters.

Option A in Table 1 illustrates the heat balance diagram of the SCM in Kiev (Ukraine). The data show a relatively low thermal efficiency. About half of the heat input is lost with flue gases, the temperature of which is 900 – 1000°C at the outlet of the recuperator. The low degree of heat recovery in this case is explained by the fact that in the initial stages of commercialization, it was decided to first concentrate on the basic technology of submerged combustion and the design of the main elements of the melter.

More recently, however, research has been carried out for a better use of the available heat in the flue gases by preheating the charging materials. Options B and C in Table 1 show the estimated characteristics of the industrial SCMs after being equipped with a bed-type charge preheater.

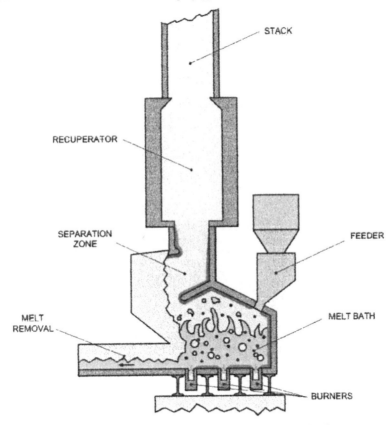

Figure 9. Schematic of industrial SCM for mineral-wool production.

Table 1. Design and performance characteristics of SCMs.

Characteristic	Options			
	A	B	C	D
Melt production rate (t/h)	2.6	3.0	4.0	8.0
Natural gas consumption (m³/h)	780	600	500	690
Combustion air consumption (m³/h)	7600	5700	4800	–
Industrial oxygen consumption (m³/h)	–	–	–	1460
Flue-gases flow rate (m³/h)	9250	6500	5400	2130
Melt temperature (°C)	1350	1350	1350	1350
Preheated air temperature (°C)	300	400	400	–
Preheated charge temperature (°C)	–	750	900	900
Flue gases in the stack temperature (°C)	940	590	180	180
Natural gas calorific value (MJ/m³)	34.3	34.3	34.3	34.3
Heat of melting (MJ/kg)	1.9	1.9	1.9	1.9
Heat input (MW)	7.4	5.7	4.8	6.6
Heat for melting (MW)	1.4	1.6	2.1	4.3
Heat losses with cooling (MW)	2.1	2.1	2.1	2.0
Heat losses with flue gases (MW)	3.6	1.6	0.4	0.16
Heat losses through walls (MW)	0.3	0.5	0.16	0.16
System thermal efficiency (%)	18.7	28.0	44.9	65.0

Option B corresponds to the existing system in Kiev, where the possible increase in production rate of the melter is currently limited by the capacity of the production line for making mats from mineral fiber. The maximum capacity of this line is 3 t/h of the melt. But even with this restriction, preheating of the charging materials increases the thermal efficiency of the melter by 50%.

The heat balances in Table 1 also show that the heat losses through the frozen melt lining are relatively high. Their impact can be reduced, however, by using larger melters, by reusing the transferred heat, and/or by using industrial oxygen. The thermal performance of the melter improves with capacity because with an increase in sizes of the melting bath, the water-cooled surface area, and consequently, the specific heat losses through the lining, per unit of manufactured product decrease. The heat transferred through the lining to cooling water is a reserve, which, if properly used, can further improve the thermal characteristics of the melter. For example, if an evaporating cooling system is used, and if the generated steam is used for process heating, energy generation, and/or for other useful purposes, then the thermal efficiency of the overall system can be further increased by a substantial amount.

4.   RESEARCH AND DEVELOPMENT OF HIGH EFFICIENCY ONE-STAGE SCM METHOD
      FOR VITRIFICATION OF HIGH-LEVEL RADIOACTIVE WASTES
      A new high-efficiency one-stage melting converter method for vitrification of High-Level Radioactive Wastes (HLRAW) has been developed and investigated. The method includes evaporation (concentration), calcination, and vitrification of HLRAW in a one-stage process inside an SCM for non-metallic minerals. Specific to the melting process is the direct combustion of a gas-oxygen-air mixture inside a melt. The experimental data for various aspects of the proposed method are presented, including SCM dimensions, burner type, data for used materials, contents of saturated salty solution and final glass product, and entrainment analysis. The effective flue-gases cleaning systems are also discussed.

4.1   Introduction

Widespread use of radioactive materials in various branches of industry, particularly in power engineering and military applications, has created a global problem for the ecological disposal of RadioActive Wastes (RAW).

For the reprocessing and disposal of RAW with high-level radionuclides, three methods can be considered: (1) reservoir storage; (2) burial in boreholes; and (3) vitrification (solidification into glass blocks) for ultimate disposal. Analysis of the recent methods of RAW localization has shown that the most reliable method of RAW storage is vitrification. This can be done by incorporating wastes into physicochemical conglomerates during glass processing from active nuclides and neutral charging materials. Vitrification allows a decrease of more than one order for RAW intended for long-term storage, and a decrease in leaching of 3 – 4 orders.

Recently the vitrification technology based on direct electrical-heating melters has been used by Savannah River Technology Center in the United States, by Mayak Processing Center in Russia and by others. Each method involves multistage processes. The disadvantages of such methods are large amount of technological equipment required for solution reprocessing (often very complicated in design and costly in operation), difficulties in equipment sealing, and hence, difficulties in the development of high-efficiency installations.

The main objective of the current work was the development and investigation of a one-stage reprocessing method for HLRAW based on their vitrification inside an SCM.

4.2   Results and Discussion

4.2.1 Basic concept of experimental equipment

An SCM for melting various non-metallic minerals was developed and investigated at the Gas Institute NASU in the late 1960s. The converter consisted of a water-cooled melting chamber with special gas burner located underneath the melt level. Combustion products penetrate through the melting materials and are removed through a flue-gases duct located in the vault of the converter. Due to its specially designed combustion process (taking place directly in the melt), the efficiency of the SCM is much greater than for ordinary bath-melting furnaces.

This type of melting converter was later used as a basis for developing a method of vitrification of medium- and HLRAW. The main features of the proposed method combine all three stages of saturated salty solution reprocessing: evaporation (concentration), calcination, and vitrification, in one continuous process in a single melting apparatus.

In details, this process is as follows: the RAW and charging materials in the form of a saturated salty solution are pumped from the top on the melt. The solution touches high temperature surface and undergoes fast evaporation with intensive spitting. Due to lighter weight, the solution will be always on the melt surface, thus preventing any possibility of vapor/steam explosions. For saturated salty-solution capacities of 360 – 900 L/h, the melt surface will not be even subcooled below the melting point. After evaporation, a solid residue mixes with melting material and melts inside it.

4.2.2 Experimental-industrial SCM "TOROS-3k"

The main objectives of this new design of converter-type glass-melting furnace (i.e., SCM), later named 'TOROS' (Thermal Treatment of Radioactive Wastes by Vitrification (in Russian abbreviation)) were to investigate the technology of a one-stage, glass-melting process from saturated salty solutions and charge; improve some design features related to glass processing from liquid solutions (primarily the SCM was designed just for granulated charge); and optimize the design of gas-burner devices.

Tests have proved the possibility of reprocessing RAW by a vitrification process in SCM-type apparatuses and directing the melt into a burial-bunker storage facility. After performing exhaustive tests with three different "TOROS"-type SCM designs, it was decided to increase the overall capacity of the reprocessing solution up to 800 L/h by using oxygen as an oxidizer. This will increase the efficiency of the whole process.

To decrease the amount of combustion products it is the best to use a gas-oxygen mixture. This will decrease the intensity of bubbling, increase the percentage of condensing gases in the combustion products, decrease the entrainment of aerosols from the first stage of the flue-gases cleaning system, and allow a wide range of vapor-gas flow temperatures over the melt. Transferring to gas-oxygen-air mixture will sharply increase the specific productivity of the SCM and its efficiency.

The scheme of the "TOROS-3k" is shown in Fig. 10.

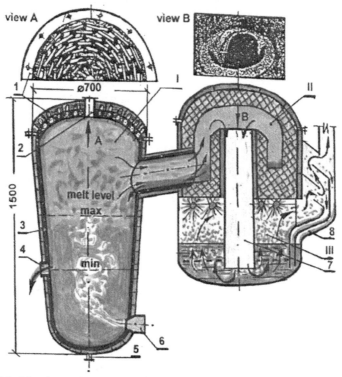

Figure 10. Principle scheme of SCM "TOROS-3k" (Pioro and Pioro, 2006): (I) SCM, (II) cyclone, and (III) bubbler, 1 – combustion products, 2 – salty-solution inlet, 3 – lining (melt-lining), 4 – tap-hole, 5 – cooling-water inlet, 6 – gas-oxygen-air burner, 7 – water-cooled flue-gases duct, and 8 – inertial precipitator.

The main features of this apparatus are:

- Gas-oxygen-air burner (Figs. 11 and 12).
- Vertical cyclone with upward flow (Fig. 10, pos. II), installed in between the SCM and the bubbler, to decrease entrainment of the molten materials and aerosols.
- New design of a flue-gases duct. Vapor-gas flow containing 60 – 75% of H_2O directed into an uninsulated duct undergoes condensation and condensate washes out all residues from the surface. This feature was tested by replacing condensation with spraying water in the upper part of the flue-gases duct.
- Sprayers were installed in the first stage of the bubbler to increase the coefficient of flue-gases cleaning.
- The vault was made elliptical shape, with specially designed thermosyphons [12] instead of pins for proper holding of lining and later melt lining (Fig. 10, view A and Fig. 13). This design helps to create a self-holding dome from the lining.

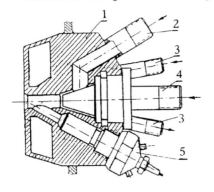 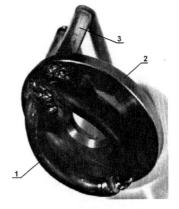

Figure 11. Gas-oxygen-air burner: 1 – water-cooled body, 2 – gas inlet, 3 – cooling water inlet, 4 – oxygen-air inlet, and 5 – igniter.

Figure 12. Photograph of the burner nozzle with two bent two-phase thermosyphons: 1 – thermosyphon evaporator, 2 – flange with nozzle for gas-oxygen-air mixture, and 3 – thermosyphon condenser.

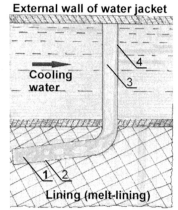

External wall of water jacket

Cooling water

Lining (melt-lining)

(a) Photograph of the vault water jacket with two-phase thermosyphons (condensers shown) (cover taken off)

(b) Vault thermosyphon: 1 – working fluid (liquid), 2 – evaporator, 3 – working fluid (vapor), 4 – condenser

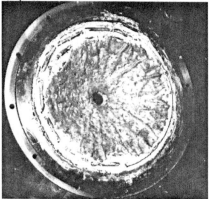

(c) Photograph of the vault with two-phase thermosyphons (evaporators shown) without lining (inside SCM view)

(d) The same as (c) with lining

Figure 13. Details of the SCM vault.

- The upward conical-shaped main body of the SCM prevents the lining from slipping down, hence to there is no need for pins. The heat-resistant lining must adhere closely to metal and be baked to solid form without significant shrinkage. Magnesite mass with 4% of Cu prepared with a formaldehyde resin (25 – 30%) corresponds to these conditions. It possesses high durability of 29.4 – 39.2 MPa.

The "TOROS-3k" apparatus has helped to test and improve such systems as:

- A system for burner remote start-up and handling of material melting in an autonomous regime. During any parameter failure the system will stop the SCM according to special procedure.
- A system for an emergency cooling-water supply, in case of a cooling system or electrical-energy-system failure.
- A system to stop the SCM without melt draining in an emergency, with subsequent restarting.

A thermal analysis was performed based on the tests with the SCM "TOROS-3k". Results of the thermal analysis are summarized in Table 2. In these tests, gas-oxygen-air mixture was used.

Calculations based on direct temperature measurements showed that at the melt temperature of 1150°C (gas consumption of 46 m^3/h), the average cooling-water temperature of 20°C (water flow rate of 22 m^3/h) and thickness of the melt lining of 15 mm the heat flux through the walls is about 75 kW/m^2 and the wall temperature is about 55°C.

Thermal analysis (Table 2) of the SCM showed that at the burner heat transfer rate of 460 kW the heat transfer rate for heating and melting of charge (useful heat) is 20 kW, the heat transfer rate for evaporation of water from the solution (useful heat) is 100 kW, and the heat loss are 240 kW (heat loss with vapor 120 kW, with flue gases 20 kW and cooling water 200 kW). Therefore, a thermal efficiency of the converter is about 26% that is usual for the water-cooled converters.

The performed calculations show that the SCM's capacity increased up to 400 L/h and for the same value of specific capacity (L/m^2 h), the thermal efficiency of "TOROS-3k" SCM increases up to 43%.

If the specific heat load (kW/m^2) increased in the SCM with Ø0.7 m to the level of the SCM with Ø0.5 m, the capacity of the new SCM would increase from 400 to 630 L/h and the efficiency up to 46%. The calculated values of the capacity and the efficiency can vary in the range of ±5 – 7% due to variations of the heat loss through the heat-resistant lining.

Further gains in the efficiency of the reprocessing could be reached by using more concentrated solutions. Accounting for the "omnivorousness" of the SCM, it is possible to melt paste-like and granulated ingredients.

The "TOROS-3k" tests were carried out with modeling-charging material consisted of datolite (2CaO B$_2$O$_3$ 2SiO$_2$ H$_2$O) – 45%, quarts sand – 15%, aluminum oxide – 18%, and soda – 22%.

The main objectives on this stage of the tests were to investigate a gas-oxygen-air burner, and determine the aerosol weight concentration in air and oxygen blowing, the dispersiveness of aerosol particles, and their sources.

The characteristics of the gas-oxygen-air burner (Figs. 16 and 17) are as follows: length 0.14 m, diameter 0.14 m, gas flow rate 60 m^3/h, and oxygen flow rate 130 m^3/h. The burner's design allows varied gas-oxygen mixing and automatic remote start-up.

The tests of the "TOROS-2" with a gas-air mixture and the "TOROS-3k" with a gas-oxygen-air mixture have shown that the aerosol concentrations were around 1.5 g/m^3 and 0.6 g/m^3, respectively.

It is known that for the same specific heat loads (kJ/m^2) on a melt surface the amount of combustion products for a gas-air mixture is 5 times higher than for a gas-oxygen mixture, and aerosol entrainment are 10 – 12.5 times higher for a gas-air mixture than for a gas-oxygen mixture.

These tests showed that the process of aerosol creation had not yet finished in the SCM. The process continued in the flue-gases duct, where the condensing origin particles (mainly high dispersion particles) were created.

The tests revealed the qualitative characteristics of the process, and some data for future developments of the SCM and the flue-gases cleaning system.

Table 2. Technical characteristics of SCM "TOROS-3k".

Characteristic	Unit	Value
Converter		
Dimensions of melting chamber: height / average inside diameter	mm /mm	1400 / 500
Minimum / maximum melt level	mm / mm	500 / 800
Melt volume	L	90–156
Capacity of saturated salty solution / Capacity of glass-RAW mixture	L/h / kg/h	200 / 100
Melt temperature in central zone (measured with optical pyrometer)	°C	1150
Average wall temperature	°C	55
Thickness of melt lining	mm	15
Heat flux through walls	kW/m^2	75
Burner		
Gas (80% propane and 20% butane) / CH$_4$ consumption	m^3/h	17.5/46
Calorific value (gas/CH$_4$)	MW/m^3	94/36
Gas pressure	MPa	0.1–0.3
Oxygen consumption (excess-air coefficient 1)	m^3/h	93/92
Oxygen pressure	MPa	0.1–0.3
Combustion products volume flow rate (gas/CH$_4$)	m^3/h	126/138
Total vapor-gas volume flow rate	m^3/h	318
Amount of condensing gases	%	78
Cooling water		
Volume flow rate	m^3/h	22
Velocity in water-jackets	m/s	0.8
Inlet / Outlet temperatures	°C	15 / 23
Heat balance		
Burner heat transfer rate (heat in)	kW	460
Heat transfer rate for heating and melting of charge (useful heat)	kW	20
Heat transfer rate for evaporation of water from solution (useful heat)	kW	100
Heat transfer rate for heating vapor (heat losses)	kW	120
Heat losses with flue gases	kW	20
Heat losses with cooling water	kW	200
Thermal efficiency	%	26

5. CONCLUSIONS

This paper is devoted to the development and design of advanced melting technologies with submerged combustion. In the submerged-combustion process the fuel (natural gas) and oxidant (air or air-oxygen mixture) are fired directly into the bath of materials being melted. The high-temperature bubbling combustion inside the melt creates a complex gas-liquid structure and a large heat-transfer surface. This significantly intensifies the heat transfer between the combustion products and the processed material. The intense mixing of the melt increases the speed of melting as well as the speed of chemical reactions of mineral formation, and improves the homogeneity of the molten product. Another positive feature of the SCM is its ability to handle relatively non-homogeneous charging materials. Based on these advantages of submerged combustion various SCM designs have been developed for producing materials for the building industry from metallurgical slag, coal slag and ash from coal-fired thermal power plants; fuming of slags of non-ferrous metals; melting silicate materials; producing mineral wool; producing molten defluorinated phosphates for agriculture; pyrohydrolysis of fluorine-containing wastes; vitrification of high-level radioactive wastes; and production of expanded-clay aggregate for lightweight concrete from non-selfbloating clays. Many of the developed technologies are intended for decreasing harmful effects of various wastes such as slags, ash, etc. on the environment by effectively reprocessing them into materials for the building industry or by safe infinite disposal of high-level radioactive wastes by including them into a glass matrix. Currently, this technology was transferred to the Gas Technology Institute in the USA.

6. ABBREVIATIONS

HLRAW	High-Level RadioActive Wastes	RAW	RadioActive Wastes
NASU	National Academy of Sciences of Ukraine	SCM	Submerged Combustion Melter

7. REFERENCES

1. Collier, Submerged combustion, International Encyclopedia of Heat & Mass Transfer, Editors: G. Hewitt, G. Shires and Y. Polezhaev, CRC Press, Boca Raton, FL, USA, 1997, pages 1105-1106.
2. Pioro, L.S., Pioro, I.L., Soroka, B.S. and Kostyuk, T.O. 2010. Advanced Melting Technologies with Submerged Combustion, RoseDog Publ. Co., Pittsburg, PA, USA, 420 pages.
3. Pioro, L.S. and Pioro, I.L., 2004. Reprocessing of Metallurgical Slag into Materials for the Building Industry, Int. J. of Waste Management, Vol. 24, No. 4, pp. 365–373.
4. Pioro, I. and Pioro, L., 2012. Development of Wasteless Combined Aggregate – Coal-Fired Steam-Generator/Melting-Converter, Proc. 20th Int. Conf. On Nuclear Engineering (ICONE-20) – ASME 2012 POWER Conf., July 30-August 3, Anaheim, CA, USA, Paper #55232, 10 pages.
5. Pioro, L.S. and Pioro I.L., 2003. Wasteless Combined Aggregate – Coal-Fired Steam-Generator/Melting-Converter, Int. J. of Waste Management, Vol. 23, No. 4, pp. 333–337.
6. Pioro, L., Osnach, A. and Pioro, I., 2003. Production of Molten Defluorinated Phosphates in an SCM, J. of Industrial and Eng. Chemistry Research, Vol. 42, pp. 6697–6704.
7. Pioro, L. and Pioro, I., 2004. High Efficiency Combined Aggregate – Submerged Combustion Melter–Electric Furnace for Vitrification of High-Level Radioactive Wastes, Proc. 12th Int. Conf. on Nuclear Eng. (ICONE-12), Washington, D.C., USA, April 25–29, Paper #49298, 4 pages.
8. Pioro, L.S., Pioro, I.L. and P'yanykh, K.E., 2002. Thermal Analysis of the One-Stage Melting Converter and Related Equipment for Vitrification of High-Level Radioactive Wastes, Proc. 10th Int. Conf. on Nuclear Eng. (ICONE-10), Arlington, Virginia, USA, April 14–18, Paper #22397.
9. Pioro, L.S., Sadovskiy, B.F. and Pioro, I.L., 2001. Research and Development of High Efficiency One-Stage Melting Converter-Bunker Method for Solidification of Radioactive Wastes into Glass, Nuclear Engineering and Design, Vol. 205, No. 1–2, pp. 133–144.
10. Pioro, L.S., Pioro, I.L. and P'yanykh, K.E., 2001. Design and Heat Transfer Calculations of Burial-Bunker for One-Stage Melting Converter for Vitrification of High-Level Radioactive Wastes, Proc. 9th Int. Conf. on Nuclear Eng. (ICONE-9), Nice, France, April 8–12, Paper #670.
11. Pioro, L. and Pioro, I., 2004. Production of Expanded-Clay Aggregate for Lightweight Concrete from Non-Selfbloating Clays, Int. J. Cement & Concrete Composites, 26 (6), pp. 639–643.
12. Pioro L.S. and Pioro I.L., 1997. Industrial Two-Phase Thermosyphons. Begell House, New York, NY, USA, 288 pages.

THE INFLUENCE OF BORATE RAW MATERIAL CHOICE ON THE GLASS MANUFACTURING PROCESS

Andrew Zamurs, David Lever, Simon Cook, and Suresh Donthu
Rio Tinto Minerals
Englewood, CO, USA

ABSTRACT

Boron brings unique benefits to many commercially important glass types, both in the production process and in the final glass product. Several options currently exist for sources of boron, ranging from minerals to pure boric oxide. For any boron-containing glass, the choice of boron source can have significant effects on several areas of glass production. Melting parameters such as energy consumption, glass formation onset temperature, and batch volume in the furnace can be affected by the selection of the boron source. In addition, furnace throughput, dust and volatile emissions and refractory corrosion may also be influenced.

This paper will describe laboratory methods used to predict these characteristics. Video techniques have been developed to observe the reaction process in its entirety. Also, methods to measure emissions of dust and volatiles, glass reaction onset temperature and reaction energy will also be discussed. Results of some of the analyses will be included and discussed.

INTRODUCTION

Borates are an essential raw material for glass making. They provide benefits to the glass forming process by way of lower viscosity, lower energy requirements, and lower glass formation onset temperature. Some of the potential benefits borates create in the finished glass are decreased dielectric constant, lower thermal expansion, increased IR absorption and increased chemical resistance.

Several different borate raw materials are commonly used in glass production. Borates are segmented into refined chemicals and minerals. In each category there are several options which vary in B_2O_3 and alkali content.

Table I. Common Borate Raw Materials for Glass Making

	Material	Formula	% B2O3
Refined chemicals			
Sodium borates	Borax pentahydrate	$Na_2O.2B_2O_3.5H_2O$	49
	Borax decahydrate	$Na_2O.2B_2O_3.10H_2O$	37
Non-sodium borates	Boric acid	$B(OH)_3$	56
Enhanced	Anhydrous borax	$Na_2O.2B_2O_3$	69
	Boric oxide	B_2O_3	100
Mineral products			
Sodium borates	Ulexite	$Na_2O.2CaO.5B_2O_3.16H_2O$	36-38*
Non sodium borates	Colemanite	$2CaO.3B_2O_3.5H_2O$	33-42*

The borate source selected for batch formulation can have dramatic effects on the glass forming process. Borate source influences the batch volume, energy consumption, furnace throughput, volatile emission rates, and the volatile species which are emitted. It is important to know these characteristics in order to have the most efficient process possible.

Several analytical methods can be combined to provide a good understanding of how borate source influences the glass forming process. This paper will detail a total of four methods to evaluate batch behavior. Two methods which are established and that are used in the glass industry are Differential Scanning Calorimetry (DSC) which can be used to evaluate the energy requirement of the glass batch, and a video of the melt which can detect the degree of batch expansion among other things. Two internally developed methods, dust emission analysis and volatile emission analysis, can be used to predict the amount of both solid and gaseous emanations from the glass forming process. The combination of these analytical techniques provides a good insight as to how borate source will influence the glass forming behavior.

ANALYTICAL METHODS:

Differential Scanning Calorimetry (DSC):
DSC was developed in 1962 and is a well-established analytical technique that is used in the glass industry to quantify the energy changes in the glass forming reactions and observe several other thermal events such as glass transition temperature T_g.

A simple comparative analysis can show how energy requirements can be affected by the borate source. By creating two glass batches designed to give the same glass oxide composition while using different sources of B_2O_3, the effect of borate source on the energy requirements of the total glass forming batch reaction can be observed. Below is an example of two LCD glass batches, Figure 1, with the sample on the left using boric acid and the sample on the right using boric oxide (anhydrous boric acid). The other raw materials are the same and the resulting glass is identical in composition.

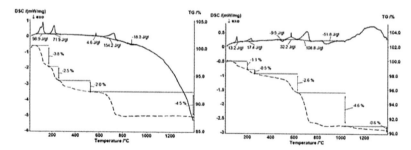

Figure 1: DSC scan of an LCD glass batch containing boric acid (left) and boric oxide (right)

When the total energies are summed, we find the boric acid batch requires 311 J/g and the boric oxide batch requires 110 J/g to reach the glass forming point. This represents a difference of over 2.5 times more energy required.

Video Melts:
This technique provides a visual depiction of the entire glass forming process in a video. If the technique is properly executed it can provide a surprising amount of information. The physical behavior of the batch can be observed. Things like batch expansion, the presence of decrepitation (the violent fragmentation of crystals caused by different levels of thermal

expansion), and puffing (the formation of low density spheres) can be observed using a video melt. For instance, Figure 2 is composed of still frames from a video that compares the performance of borax pentahydrate (5-mol) to that of anhydrous borax (AB) in an insulation fiberglass formulation. In the video the batch expansion in the 5-mol samples is very visible; the batch approximately doubles in size at one point during the observation period.

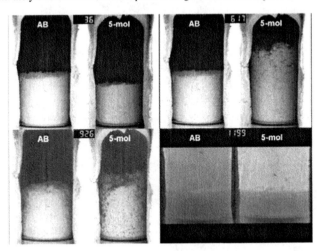

Figure 2: Visual comparison of AB to 5-mol in an insulation fiberglass

In addition to the physical batch behavior, other information can be gathered as well. The temperature at which the batch begins to collapse can easily be seen, and also the batch free time can be determined and a comparative analysis can be made between different raw materials. When the whole glass forming process is observed from start to finish it provide a good idea of how a potential borate source can affect the melting process.

Dust Emission Analysis:

As seen in the video melt section, borate source can have a dramatic effect on the batch behavior. Some of the phenomenon, namely puffing and decrepitation, which are observed in the video melt, also have a pronounced effect on the dust emission of a batch. There are several raw materials that will contribute to dust emission, these include, borax pentahydrate, colemanite, and dolomite.

In order to quantify and compare the effect of borate source on the dust emission of a batch the apparatus seen in fig 3 was devised. Initially an air flow is created through the furnace which is set at a given temperature. A sample is then inserted into the furnace the emitted dust is collected. The air flow is adjusted so that only the emitted dust due to batch behavior is collected.

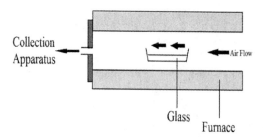

Figure 3: Dust collection apparatus

Initial experiments show that differences in dust emission can be observed between glass batch containing different borates sources. The experiments were conducted using the same experimental conditions on an insulation fiberglass formulation designed to give the same glass oxide composition, with the only difference being the borate source.

Table II: Dust Emission results, in % of batch emitted

Borate	Test 1	Test 2	Test 3	Mean (% of sample)
Anhydrous borax	0.074	0.071	0.092	0.08
Borax pentahydrate	0.84	0.86	1.00	0.90

This difference in dust emission is due to the fact the borax pentahydrate demonstrates a behavior referred to as "puffing" in which the material acts similar to popcorn as it is heated, creating small low density spherical particles and in the process creating airborne dust. Anhydrous borax does not experience this phenomenon because in this material the chemical water has been removed. In this case the dust emission varies by an order of magnitude solely due to the borate source.

Volatile Emission Analysis:
Volatile species of a glass batch are a well-known. The raw materials of the batch, the heating cycle and several other factors affect the species that are emitted and the rate at which they are emitted. Borates are some of the more volatile components in glass batch. The ability to measure boron volatility in a lab setting is not widespread. For this work, a system to collect and analyze volatile emissions was constructed and validated (Figure 4).

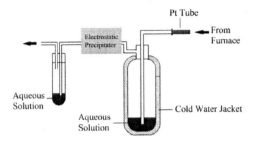

Figure 4: Volatile Collection Apparatus

The sample to be analyzed is placed in the furnace. Air is drawn through the tube furnace and through the collection apparatus. Volatile species are entrained in the air steam and deposited in the collection system. Once the sample has completed analysis the resulting solution is then analyzed by ICP-OES to quantify the total amount of volatiles emitted.

In order to determine the efficacy and reproducibility of the system several different experiments were completed. The initial concern was the efficacy of the system: especially any volatile emissions that would be collected. In order to test how effective the system was at collecting volatiles, anhydrous borax was run as a standard, and a mass balance was completed, with results shown in Table III.

Table III: Collection Apparatus Efficacy Analysis

	Na mg	%	B mg	%
Initial Wt mg	1823	100	1726	100
Final Wt mg	1698	93	1651	96
Collected	87	5	43	2
Sum	1785	98	1694	98

Calculation of the mass balance determined that the system has an efficacy of 98%. These results validate the system and allow further experiments to be completed with confidence in the results. Next the reproducibility of the system was determined. In order to determine the reproducibility an internal glass standard was created and tested several times, with results for sodium and boron shown in Table IV.

Table IV: Reproducibly of the volatile analysis

	mg element per g glass	
Repeat	B	Na
1	2.9	5.3
2	2.9	5.4
3	2.8	5.1

The results show the method has good reproducibility. With the method validated and the reproducibility confirmed some preliminary experiments were done.

In order to better understand the relationship of volatility and temperature a simple experiment was designed and completed. The results are shown in Figure 5. The plot shows that the volatility displays Arrhenius behavior under these conditions.

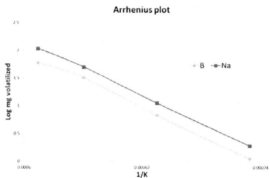

Fig 5: Arrhenius Plot of Na and B volatility compared to temperature

A second parameter that was investigated was the amount of volatility over time. The results are presented in fig 6.

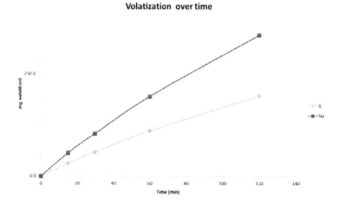

Figure 6: Volatility over time for B and Na

The relationship between volatility and time is not linear. The cause of this is not clear and needs to be investigated further, but it is likely the there is some sodium and boron depletion of the surface of the glass melt over time. The changing concentration at the glass surface over time could affect the volatilization rate.

At the time of this publication, not enough data has been gathered to quantify how borate source affect volatility of a glass forming process.

Future Work:
The dust and volatile emission analyses offer a wealth of information and much of it still needs to be explored. For dust emission analysis the next stage of development will include completing the analysis for all types of borates, and also further investigation on how temperature affects the dust emission of a given batch.

There is even greater potential for further research regarding the volatile emission analysis. The immediate investigation will determine the effect of atmosphere composition on the volatilization rate, and breaking down volatility into temperature ranges to determine at what point the majority of volatile losses occur.

CONCLUSION:
Many parameters of the glass forming process can be affected by the borate source. Using several established methods, DSC and video melts, and some internally developed analyses, dust and volatile emission analysis, many different features of a glass melt can be observed. When these analytical techniques are combined a good understanding of borate material behavior in glass batch and glass melts can be obtained.

ELECTRICAL HEATING SYSTEMS FOR MELTING TANKS, FOREHEARTHS, AND FEEDERS

Werner Bock and Gunther Bock
Bock Energietechnik
Floss, Germany

David Boothe
Allstates Refractory Contractors, LLC
Toledo, OH

INTRODUCTION

We present a brief review of electric heating (melting and boosting) concepts and equipment for the glass manufacturing industry. The public main electric supplies are built as three phase systems. Based on geometrical reasons, which exist in glass furnaces, three phase systems cannot be used every time. Therefore, it is necessary to burden the main supply symmetrically so that ideal requirements are available for glass melting and heating.

The public mains supplies are built as three phase systems.

Three phase systems can't be used every time. So it is necessary to burden the mains supply symmetrically for glass melting.   2

1) SINGLE PHASE SYSTEMS

The simplest method is a single phase system load through a pair of electrodes. This is most often used for throat and small area, concentrated heating applications. However, this arrangement is only suitable for a limited demand because of phase symmetry. This depends on the local conditions or rather which supply unbalance is allowed by the energy supplier.

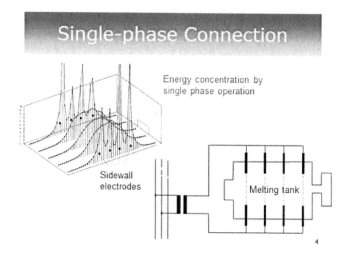

The same is also true during the operation of two single phase transformers connected to the Alternating Current net (a so-called V-connection). The alternating current of the first transformer is shifted 120 degrees from the second transformer. The results can be seen in the diagram showing an expanded distribution of energy density present in the glass bath.

Here too, the common phase has a load 1.73 times greater than the other two phases. Additionally, this is the upper unbalance limit generally accepted by the power suppliers.

For larger power demands utilizing single phase operation, it is possible to convert from a three phase grid to single phase output by using a Balun connection. A Balun is type of transformer used to convert an unbalanced signal to a balanced one or vice versa. Baluns are used to isolate a transmission line and provide a balanced output. The term Balun is a contraction of balanced to unbalanced transformer connection.

A capacitor group and a choke group are connected between each of two of the incoming phases and the actual load is taken up by the third phase which is the electrode system.

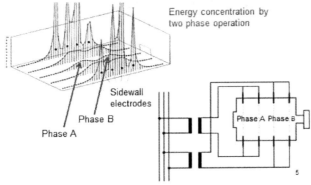

2) MULTIPLE PHASE SYSTEMS

Three phase systems require an extensive symmetrical distribution of phases and resistances. They are therefore most suitable for hexagonal melter designs.

A variety of circuit designs are available, such as the delta or the open delta circuits.

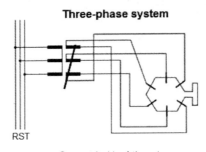

In addition, it is also possible to apply three single phase systems to a melter layout. They can be loaded differentially in different zones to achieve a desired energy distribution in the melt tank.

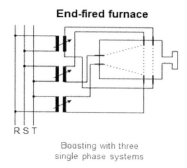

Multiple phase systems

End-fired furnace

R S T

Boosting with three
single phase systems

9

The power distribution achieved by a "Scott" switch should be mentioned as the third possibility. In this configuration, the three-phase AC connection is changed to two single phase power sources. Because both phases are offset electrically by 90°, this method is especially suitable for square tanks or electrode arrays.

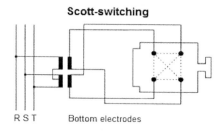

Multiple phase systems

Scott-switching

R S T   Bottom electrodes

„Scott systems"
Symmetrical system load

10

3) MODELING METHODS
Two modeling techniques are primarily used for design of installations.

a) Physical modeling
Physical modeling is performed in an appropriately scaled Plexiglas model with a modeling fluid matched to the specific glass characteristics.

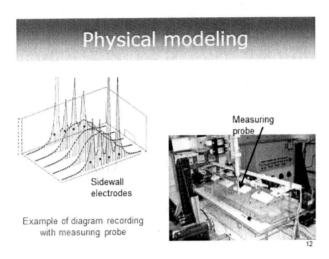

Example of diagram recording with measuring probe

This modeling method provides a sound engineering approach and very good graphic results. However, it is very time-consuming and is therefore quite costly. This method can be used to illustrate the mathematical model results.

In other words, the results of mathematical modeling can be shown in greater detail in the physical model. Both models should be reasonably consistent in their results.

b) Mathematical modeling

Mathematical modeling offers the advantage of obtaining the results faster. However, mathematical modeling is a very theoretical approach, whereas physical modeling is closer to reality and gives a more perceptive impression. At the beginning of a mathematical modeling, theoretical approaches that are a result of practical experiences must first be clearly identified and applied.

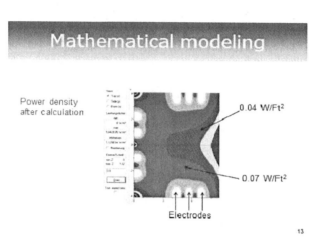

Based on these experiences, variations such as electrode positions, number of electrodes and various throughput loads can be entered in the computer program. From these results, the best conditions for the application of electric power supply can be determined. This method has been performed at many plants and compared with the results of the physical models, as well as with actual values measured on operating furnaces. This has allowed us to collect meaningful, reproducible results.

Advantages of Side vs. Bottom Electrodes

Basically, it is assumed that the side electrode position form a horizontal thermal barrier in the furnace. On the other hand, the bottom electrode position forms a vertical barrier.

The side electrode position is useful if one wants to supply energy at a given height in the melt bath. (i.e. most of the energy is released at or above the electrodes. With bottom electrodes, energy is supplied over the entire height of the insertion depth. Bottom electrodes are particularly useful in boosted tanks where quality improvement (homogenization) is desired in the glass mass.

In all electric melters, the application of electrodes is varied dependent on the furnace construction and type of glass. Bottom, side and even over the top electrodes are in use as combinations.

4) POWER SUPPLIES (TRANSFORMERS AND CONTROL)

The conversion of power from the main supply to the electrode voltage and the associated control (adjustable output) can be performed by many ways and means. Some suggestions are:

a) Step transformer
Here, a secondary winding is provided with manual stepped taps. The supplied power is varied by changing the taps. The disadvantage is that only coarse steps can be taken and then only with the power shut off.

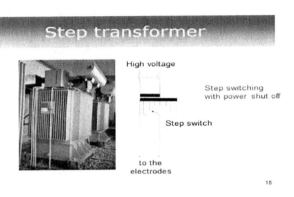

b) Variable transformers
Voltage control is accomplished with an AC transformer by means of a stator and a rotor. As a result, a given input voltage can be infinitely varied to produce an output voltage under power. Somewhat higher losses result from the combination of a fixed and a rotary transformer. This condition results in conversion losses in the form of magnetization and copper losses due to the combination of the transformer systems.

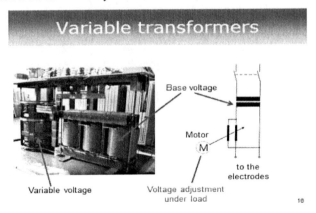

c) Control transformers

With these transformers, the secondary voltage is tapped by means of a wiper contact. The core rotation drives the copper contacts along the secondary windings, allowing a step-less voltage variation. This sliding contact provides a continuous connection between the copper slide path built in the core, and the secondary windings groove. This design provides a large voltage range, which can be adjusted under load.

Advantages of this system are:
– Low system losses
– Small footprint

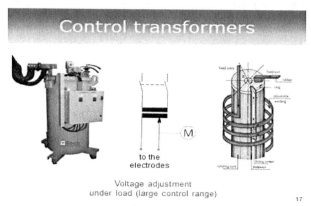

Control transformers

to the
electrodes

Voltage adjustment
under load (large control range)

d) Voltage control with series transformers

In this method, a control unit and a series transformer using a defined voltage range is added or subtracted. This method is particularly useful with step transformers or fixed transformers to enable adjustment over a small voltage range. It is also useful for additions to existing systems for automatic voltage control.

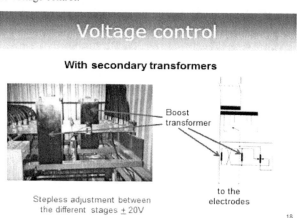

Voltage control

With secondary transformers

Boost
transformer

to the
electrodes

Stepless adjustment between
the different stages ± 20V

e) Thyristor – transformer systems

Power control with thyristors is an inexpensive method. A thyristor is a four layered semiconductor that is often used for handling large amounts of power. While a thyristor can be turned off and on, it can also be used to regulate power using phase angle control. This allows the amount of power to be controlled by adjusting the angle of the current input. A good example of this is a dimmer switch on a light.

However, it should be noted that the operating voltage is only about 75% of the modulation of the thyristor. It is therefore often necessary to utilize a matching transformer with several taps with which the necessary voltage can be matched to the electrodes. In certain cases, it is recommended to connect a second thyristor in parallel, which produces a further increase in voltage (so-called voltage sequence control).

In our opinion, the control transformers are the most robust and reliable solution for the glass industry. In the severe operating environment, where dust and fumes exist, sealed oil transformers have proven themselves.

When costs are compared, the control transformer is competitive because in a single unit the power supply and control transformer exists.

5) ELECTRODE HOLDERS AND ACCESSORIES
We distinguish between two main groups of electrode holders.

a) Open electrode holders (open cooling circulation)
Commonly referred to as "splash" type holders, these are used primarily in side mounted installations where accessibility is easy. They are insensitive to slightly contaminated water, but there are frequent problems with water drainage and electrode corrosion. In addition, they use more water, because of evaporation.

Electrode holder with open cooling

21

b) Electrode holders with a closed cooling system
Electrode holders with a closed cooling system have proven themselves very well in practice in a number of cooling designs. They work largely without any problems; however it is imperative to pay extremely close attention to the cooling water quality. Water treatment and water recirculation systems are required to prevent build up in the internal cooling chamber of the electrode and reduce cooling costs.

Electrode holder with closed cooling

22

c) Electrode support systems

We have developed a device for mounting and support primarily for bottom electrodes, which allows the electrodes to be safely pushed in and secured. Through coarse and fine settings, precise adjustment of the support system is assured. This device can also be conveniently used for installing an electrode extension. The use of a welded plate enables easy positioning of the device. The distance between the electrode and support is freely selectable since the support insulation can be freely moved and rotated by 360°.

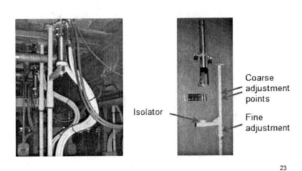

Electrode support systems

Isolator — Coarse adjustment points / Fine adjustment

23

d) Electrode connection clamps

Current contact is achieved with a nickel-plated copper clamp. It can be supplied in different versions as indicated in these diagrams. The current load on the standard clamp is 2500A. The clamping screws are equipped with spring washers, so that the contact ability is assured even with temperature fluctuations.

Electrode connection clamps

Max. Current: 2.500A

24

6) POWER SUPPLIES:

The supply of power between transformers and electrodes can be subdivided into two methods.

a) Cable systems

Due to the high current loads, a number of individual cables are run in parallel. This can cause a problem of different current loads in parallel cables, resulting in electromagnetic fields. For this reason, we have developed an installation system which avoids large differences in the individual cables.

An optimum current load is achieved by the following points:
– Defined cable distances
– Combination of supply and return lines to the electrodes
– Consideration of the various current phases

Cable systems

Power supply by cable systems

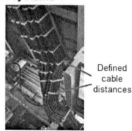

Defined
cable
distances

Possibility of laying in
constricted rooms

26

b) Bus bar systems

Bus bar systems are well suited for straight, usually longer current supply distances (usually longer than 50 – 60 feet). In this case they offer the following advantages:
– Space relationships allow long symmetrical runs
– Higher ambient temperatures are involved

The decision for selecting copper or aluminum designs depends frequently on available space, ambient temperatures and material prices.

Bus bar systems

Power supply by bus bar systems

Compact energy supply systems
well suited for straight, usually longer current
supplies in constricted rooms

27

7) ELECTRODE COOLING

Electrode cooling is an often neglected topic with electrical heating systems. Many problems with electrode holders can be traced to insufficient cooling or inadequate cooling water quality. It is therefore imperative to maintain the cooling water hardness as well as the pH value within certain limits. (40 - 50mg CaO/l - total chloride content: < 50mg/l)

Likewise, the cooling water temperatures must be kept between the limits of 100°F and 130°F. In addition, an uninterruptible supply of cooling water must be assured. This concerns circulation pumps and heat exchangers in a redundant system and/or an uninterruptible cooling water supply during power outages by activation of city water or water tower storage.

Electrode cooling

Complete cooling water system with

- heat exchanger [1]
- circulation pump [2]
- tank [3]
- control cabinet [4]
- emergency water [5]
  supply

29

It is a further recommendation that the cooling water returns be monitored with regard to temperature and flow rate.

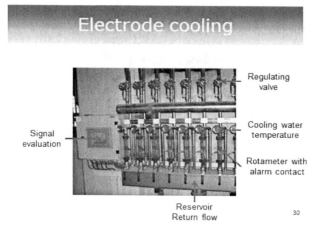

8) FOREHEARTH HEATING

a) Heating systems

Heating with electrodes in the forehearth is used primarily for temperature equalization across the forehearth cross-section. Depending on the type of glass and the feeder and forehearth size, a combination of 8 to 12 electrodes can be installed. Power is controlled depending on the temperature gradient, whereby each side is independent from the other. Power adjustment is usually performed by thyristor control.

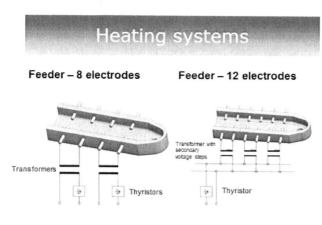

b) Electrode shapes
Electrode shapes are dependent on the glass composition and whether cooled or non-cooled electrodes are used.

The electrode (pictured at the top can optionally be used as an air-cooled or uncooled design.

The water-cooled electrode, pictured in the middle is mostly used for glasses that are not very prone to crystallization.

The electrode pictured at the bottom has a ceramic coating. This serves as protection against corrosion in the area through the refractory block.

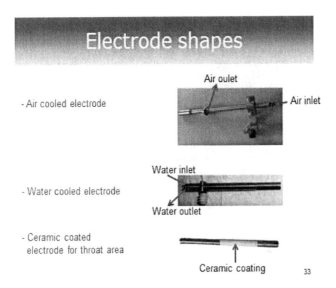

9) INDIRECT FEEDER BOWL HEATING
A heating conductor design, consisting of flat band elements formed to the contour of the feeder bowl, produces a maximum thermal contact surface on the outside of the bowl. Through this simple construction, a significant improvement in temperature homogeneity and thereby product quality can be achieved. In practice, this heating application compensates for the heating losses which would otherwise occur in the feeder bowl. The power requirements range from 1 - 5 kW, depending on size of the bowl.

This type of heating is particularly advantageous with:
   –Dark-colored glasses (better glass homogeneity in the article)
   –Production of small items (vials and peanut ware)
   –Low shear-cut production (large glass blocks and lenses and glass   insulators

Specific Advantages:
1. Temperature compensation at the feeder outlet and equalization in the bowl. Significant effects have been noted in multi-gob operations due to better temperature equalization across the orifice.
2. Shortened heat up time after a bowl or orifice change. Production downtime is shorter.
3. In all electrically (electrode) heated forehearths, it serves as a replacement for surface heating.

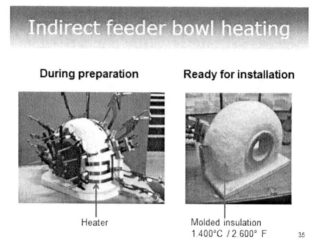

CONCLUSION

We have covered a quick review of the electric boosting and melting equipment and design from our prospective. There are certainly significantly more criteria for the choices of location, power requirements, and control. The particular application and customer desired benefits must be provided and closely studied to provide the most efficient and cost effective system.

We would like to thank the GMIC and the Board of Review for the opportunity to present this paper. We appreciate your attention and look forward to your questions. Thank you.

A FURNACE COMBUSTION SYSTEM CONVERSION BY FLAMMATEC DURING
OPERATION AT LIBBEY, INC. - HOW FLAME GEOMETRY IMPROVEMENT AND
EXCESS AIR CONTROL CONTRIBUTES TO FUEL SAVINGS

Dan Cetnar, Libbey, Inc.
Petr Vojtech, Flammatec, Inc.
H.P.H. Muijsenberg, Glass Service Inc.

ABSTRACT

A combustion system conversion was accomplished at the Shreveport, LA, plant of Libbey, Inc., during which a steady furnace operation was maintained. Since the furnace operation was stable, a direct comparison could be made to the previous operation and evaluated versus the new burner system installation. New burners were installed by FlammaTec, Inc., which is utilizes a dual natural gas flow control system with truly independent gas flow control for each of the two (2) gas streams. By improving the flame shape control, and at an identical glass quality, an improvement to the stoichiometric combustion ratio was achieved, yielding some fuel savings. The conversion of the combustion systems will include a comparison of some mathematical modeling, as well as some illustrations of the flame imaging.

This paper will present the practical results of an application of a flexible duel gas injector by FlammaTec, Ltd. (FlammaTec), at the tableware plant for Libbey, Inc. (Libbey), in Shreveport, Louisiana. In fact, the application of the FlammaTec Flex Burner was applied to two (2) furnaces at the Libbey Shreveport plant.

In today's world market there are many issues facing the glass manufacturing industry, including both economic and environmental challenges. The price of energy is high, especially for fuel oil, and although currently the price for natural gas is low in the United States, it is unlikely to remain low over an extended period of time. Also, there are many environmental regulations in terms of emissions levels, and the related potential penalties associated thereto, if these limits are not achieved.

Libbey sought to replace its existing burners and achieve an improved furnace operation and energy savings.

Existing or current burner technology has utilized a duel gas injector since the 1960's. Gaz de France published its first version of this twin gas burner in 1968[1]. Subsequently, other burner manufactures introduced their own versions of the dual gas nozzle burners, but their identifying feature was that there was one (1) gas inlet to the burner with the independent gas streams only later separated within the burner itself.

Advantages of a new burner concept were developed for the FlammaTec Flex Burner in the years 2006 and 2007. The Flex Burner utilizes two (2) fully separate gas inlets and gas flows into the burner which would be controlled and measured independently. The burner tip was optimized utilizing computational fluid dynamic modeling to minimize turbulence at the burner tip. The burner nozzle was also designed to be fully adjustable. Hence, the new burner design offered some technological advantages.

Figure 1 shows an illustration of the FlammaTec Flex Burner with the duel gas injector that utilizes precision adjustments of the inner nozzle to have precise flame shape control.

Figure 1

Figure 2 illustrates how the Flex Burner design was optimized utilizing computer modeling to minimize flame turbulence. This figure shows how the development of each burner nozzle was optimized. Higher natural gas turbulence is illustrated in red, while the lower turbulence is shown in blue. The FlammaTec Flex Burner optimizes each of the nozzles used for different applications and different capacities.

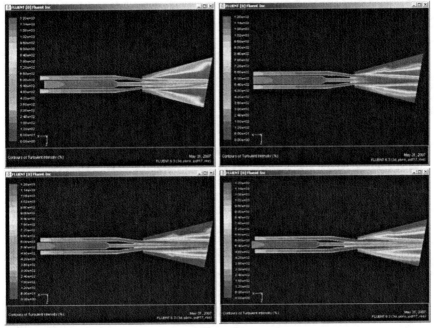

Figure 2

The specific burner nozzle dimensions are measured and modeled using computation fluid dynamic models to minimize flame turbulence as shown in Figure 3.

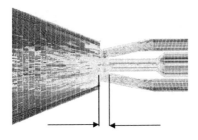

Figure 3

Figure 4 shows the actual installation of the FlammaTec Flex Burners on Furnace "C" at Shreveport.

Figure 4 [2]

The goal for Libbey was to investigate and evaluate any potential benefits resulting from the use of a dual gas nozzle, low NOx burner with independent flow control through both the primary and secondary fuel streams. The criteria for the burner application were to maintain a stable furnace operation within Title V permit limits, and with no negative impact to furnace emissions, glass quality, or the melting energy.

Prior to the application of the FlammaTec Flex Burners, Libbey had tried several different types of burners over the recent years in an effort to improve efficiency, each with varying degrees of success. The most recent improvement prior to the FlammaTec burners was used as the furnace operating baseline for comparison purposes. Essentially, the FlammaTec Flex Burners could be evaluated quite studiously, since the Libbey experience with combustion control upon their furnaces has provided them with very specific understandings of their limitations as it pertains to their furnace emission rates, glass quality, furnace efficiency, and glass redox.

Libbey established the trial procedures for the FlammaTec Flex Burner, including the simple goal of changing the burners out port-by-port over a six (6) to eight (8) week period without negatively impacting any of the primary furnace operating criteria. After each port conversion of the burners, the furnace was observed for at least a one-week period prior to the next port conversion. Note that the fuel distribution was not changed from the previous operating conditions for the furnace.

The first application of the FlammaTec Flex Burners was applied to Furnace "C" at the Libbey plant in Shreveport, Louisiana. This furnace has been continuously operated since 2007, producing high quality, low iron, soda lime glass for tableware production, utilizing a carbon/sulfate fining system. Furnace "C" is a five (5) port cross-fired furnace with one (1) burner per port and it is equipped with independent flow control on each port. The pull rates for the furnace were typically between 150 and 190 US tpd, and the furnace would usually run this tonnage within a +/- 10 US tpd pull rate for months at a time. The cullet ratio on the furnace was twenty eight percent (28%). Furthermore, the furnace has multiple oxygen sensors installed that are used for continuous monitoring and control of the furnace atmosphere.

Figure 5 shows the first installation of the FlammaTec burners on Furnace "C" at the Shreveport plant of Libbey in March, 2011.

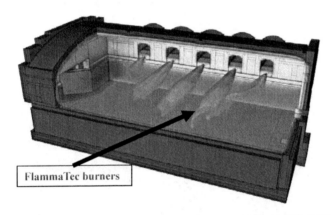

FlammaTec burners

Furnace type:	Five ports; cross fired
Glass type:	Soda lime
Pull rate:	160 MTPD
Cullet:	28 %
Total gas consumption:	xxx Nm3/h
Electric boosting:	Not operated

Figure 5

Figures 6 and 7 illustrate the flame development on Port Numbers three (3) and five (5) on Furnace "C". Note the nice development of the flame envelope with increased soot development in the central region of the flame. This low velocity and high-luminosity flame yields improved furnace efficiencies, as well as reduced NO_x emissions.

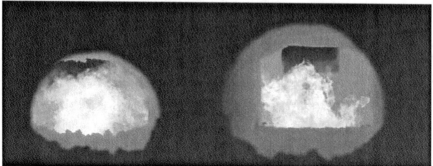

Figure 6 [3]   Figure 7 [4]

Figure 8 shows the furnace efficiency on Furnace "C" from the previous burners to the new installation of the FlammaTec Flex Burners. The figure shows the different tonnages upon the furnace, with the operating values in mm Btu/ton. Each division upon the "Y" axis is gauged in 0.5 mm Btu/ton increments. There was a 2.5% fuel savings on Furnace "C" with the direct conversion of the FlammaTec burners from the previous burners. There were no other changes made to the furnace at this time.

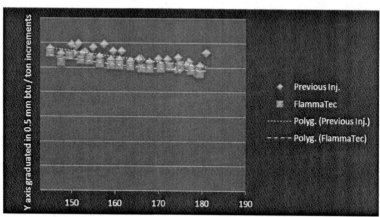

Figure 8

After the successful application of the FlammaTec Flex Burners on Furnace "C" at Libbey, they decided to install these burners upon a second furnace at Shreveport, Furnace "A".

Furnace "A" has been continuously operated since January 2012, also producing high quality, low iron, soda lime glass for tableware production, utilizing a carbon/sulfate fining system. Furnace "A" is a five (5) port cross-fired furnace, which is electrically boosted and utilizes one (1) burner per port, with independent flow control to each of these ports. The typical pull rates are between 160 and 230 US tpd. The cullet ratio is 18% to 28%.

Furnace "A" was converted to the FlammaTec burners during a routine furnace rebuild in December 2011. Figure 9 shows the improved furnace efficiency on Furnace "A" with the new burners. The furnace tonnage is shown with the corresponding energy usage in 0.5 mm Btu/ton increments for their respective tonnages. It should be noted that although the FlammaTec burner contributed to the overall furnace efficiency improvements, the furnace was also rebuilt at this time with many other improvements made concurrently.

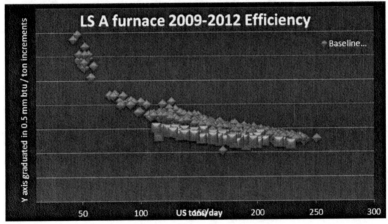

Figure 9

Libbey summarized the transition to the FlammaTec burners as occurring "as planned" with no abnormal consequences. The initial burner set-up was made by visual observation of the flame shape and length to most closely replicate the baseline conditions. Secondary air / fuel ratios varied between 10% and 20% and were also set visually, based primarily on flame geometry. It quickly became apparent that the flame envelope stability was much more consistent than with the previous burners.

Another important furnace operating parameter for Libbey was the glass redox, which is the relationship between ferric and ferrous iron. Libbey knows from experience that the combustion atmosphere has a considerable influence on this value. The improved stability of the FlammaTec burners became an advantage to the stability of the redox. Based upon previous experience the net oxygen level in the furnace atmosphere was decreased in several steps from 2.5% excess oxygen to 1.5% excess oxygen. Historically, this change caused a measurable shift in glass redox, negatively impacting glass color.

The adjustment with the FlammaTec burners in net oxygen level did not cause a shift in glass redox that had been previously seen. In fact, since the FlammaTec burner installations were made, the overall glass redox has become slightly more stable than historically available. See Figure 10 which illustrates the improved stability of the glass redox values upon both Furnaces "C" and "A" at Libbey.

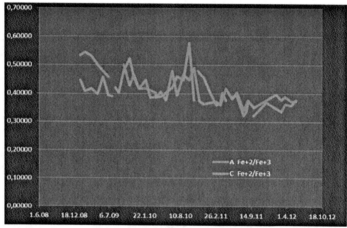

Figure 10

In summary, the use of the FlammaTec Flex Burners enabled the melting operation to be adjusted to run with a lower oxygen content without negatively impacting glass color. The sustained operation at a lower oxygen content results in an improvement in melting efficiency on an energy per ton basis. The average reduction over the entire operating range was in the range of 2.5% mm Btu/ton

The FlammaTec burners have provided a more stable combustion process that was unachievable with previous burners, resulting in the furnace operation stability, and a positive shift in the glass redox. Note that burner adjustment throughout the operating range has been improved from previous burners.

Libbey did note that there was an increase in burner cleaning frequency with the FlammaTec Flex Burners, primarily on Port Number 1, since as the burner tips collect dirt, the impact on the flame is more significant than on previous burners. Additionally, with the natural gas and cooling piping requirements (two supply lines each), there was an opportunity for more physical connections and fittings, thus creating increased leakage potentials.

The second installation of the FlammaTec burners on Furnace "A" was implemented with separate gas supply lines to each burner and with continuous flow measurement on both the primary and secondary fuels. This enabled Libbey to remotely and independently control each of the gas flows to the burners from their control room.

For reference, Figures 11 and 12 show a conversion on a container furnace with the FlammaTec Flex Burners. Note the fuel efficiency improvement from 3.66 mg joule/kg (x mm Btu/ton) to 3.54 mg joule/kg (x mm Btu/ton) or an improvement of x % furnace efficiency.

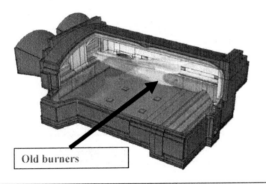

Old burners

Furnace type:	Container
Glass type:	Green
Pull rate:	251 MTPD
Cullet:	53 %
Total gas consumption:	1,005 Nm3/h
Heat value of gas:	11,192 kcal/kg
Combustion heat energy:	9,550 kW
Electric boosting:	1,082 kW
NOx:	920 mg/Nm3 (8% O2)
Melter area:	82.8 m2
Glass depth melter:	1,150 mm
Specific pull:	3.03 MTPD/m2
Specific energy consumption:	3.66 MJ/kg

Figure 11

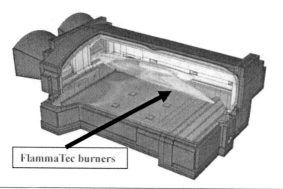

Furnace type:	Container
Glass type:	Green
Pull rate:	251 MTPD
Cullet:	53 %
Total gas consumption:	968 Nm3/h
Heat value of gas:	11,192 kcal/kg
Combustion heat energy:	9,201 kW
Electric boosting:	1,082 kW
NOx:	650 mg/Nm3 (8% O2)
Melter area:	82.8 m2
Glass depth melter:	1,150 mm
Specific pull:	3.03 MTPD/m2
Specific energy consumption:	3.54 MJ/kg

Figure 12

Figure 13 also shows a float furnace installation with a very efficient furnace with only 5.18 mg joule/kg (x mm Btu/ton).

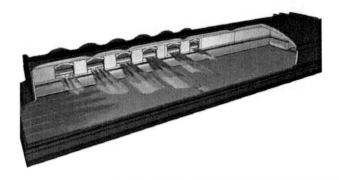

Furnace type:	Float
Glass type:	White
Pull rate:	700 MTPD
Cullet:	30%
Total gas consumption:	4,300 Nm3/h
Heat value of gas:	10,910 kcal/kg
Combustion heat energy:	41,939 kW
NOx:	< 1000 mg/Nm3 (8% O2)
Melter area:	522 m2
Glass depth melter:	1,325 mm
Specific pull:	1.34 MTPD/m2
Specific energy consumption:	5.18 MJ/kg

Figure 13

In conclusion, the FlammaTec Flex Burner has truly independent gas flow adjustment for both the inner and outer burner nozzles. Thus, the burner offers improved flame shape control, and therefore tighter flame stability, which yields more flame luminosity and therefore improved furnace operational performance. The improved flame shape control, and therefore excess oxygen levels within the furnace can yield improved energy efficiency.

FlammaTec expresses its sincere appreciation to Libbey, Inc. and its Shreveport, LA, facility which have made this presentation possible.

REFERENCES

(1) Gaz de France, 1968, Twin jet burner nozzle
(2) Libbey, Inc. photograph of the FlammaTec burner installation at the Shreveport, LA facility
(3) Libbey, Inc. photograph of the FlammaTec burner installation, Port Number 3, at the Shreveport, LA facility
(4) Libbey, Inc. photograph of the FlammaTec burner installation, Port Number 5, at the Shreveport, LA facility

Coatings and Strengthening

COLOR CONTROL OF GLASS AND MULTILAYER COATINGS ON GLASS

David Haskins, Paul A. Medwick, and Mehran Arbab
PPG Industries, Inc. Glass Research and Development

ABSTRACT

With careful control of coloring oxides in the production of float glass, and precise control of the layer thicknesses in multi-layer coatings, aesthetic issues on building facades, such as checker boarding and color change with viewing angle can be minimized. Internationally-accepted systems have been developed for describing and quantifying color numerically, which, when combined with knowledge of the function of coloring oxides in glass, can provide the glassmaker with valuable tools for managing the appearance of glass products. Similarly, with an understanding of the interactions between coating layers, consistency in the appearance of multi-layer coatings on glass can be achieved. This paper will address the color phenomena in architectural glass and guidelines for controlling color in glassmaking and coated glass.

INTRODUCTION

We understand our environment through our five senses. The most discriminating judge of a building is our vision. Other senses, including touch (sensation of heat) and hearing (acoustics) can also influence our appreciation of a building. However, our first and longest lasting impressions are developed through our impression of the aesthetics of a structure, which in turn is defined by its form, texture, and color. The architect is keenly aware of this and appearance is at the heart of the design. Perhaps form is dominant; while texture and color serve the form; they can at times be more visible than their master. This is particularly true if the architect chooses to use vivid colors and prominent textures, or when unintended attributes are incorporated in the building materials.

Regarding window glass, color issues might include off-color product, non-uniformities in the product, or variations in color with angle of view. The human eye is highly sensitive to color difference. Uniform off-color glass panels can become troublesome if there is significant change in hue. For example, if green-blue glass is specified, yellow-green glass might be perceived as out of place, while most observers might not object to a moderate deviation from specification within the green-blue family of colors. On the other hand, the eye readily recognizes relatively small differences in color between neighboring windows or curtain wall panels. This is especially important in the case of the curbside observer. Looking at a building, the observer has a global view and can see variability in a single look. The building occupant, on the other hand, is most likely looking through one window at a time. A non-uniformity problem is even more pronounced on the curbside as the observer is looking at, instead of, through the glass, and reflected color variations will amplify the perceived difference. Therefore, off-specification coated glass can be more problematic than uncoated glass. The subject of coated glass leads to the third source of unintended color issues, which is the variation of color with angle. Again, this variation can manifest itself as either objectionable uniform hue at angle (e.g., reddish cast) or side by side variability (e.g., different shades of blue).

The problem of glass color is of course not limited to architectural applications. Similar problems may exist in bottling, tableware and even in insulation materials. The science of glass and coated glass color is advanced, and the industry has developed significant proprietary tools in color control. In this paper, we will provide an overview of the basic physics of color in tinted glass and thin film coated glass and guidelines in color design and control.

BACKGROUND

Controlling color of float glass and coated glass requires an understanding of how light interacts with the glass, methods for measuring that interaction, and systems for translating the measurements into meaningful descriptions of color and color acceptability.

Figure I is a depiction of the basic interaction of incident light with glass. Incident light on glass (I_0) is either transmitted (I_T) through the glass, reflected (I_R) from its surfaces, or internally absorbed (A).

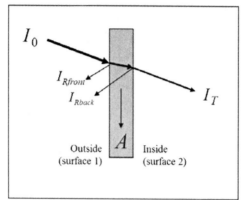

Figure I – Interaction of incident light with glass

These interactions can be measured with a spectrophotometer to produce spectra that are a function of wavelength. Each glass or coated glass color exhibits a unique transmittance or reflectance spectrum that is representative of the unique optical design of the product. Examples of transmittance and reflectance spectra, at near-normal incidence, for a typical coated glass are shown in Figure II. The figure shows that the reflected spectrum, and therefore color, of a coated glass depends on the surface that is measured. On the other hand, for an uncoated glass, reflectance of both sides would be primarily the same. An exception to the latter exists in the case of float glass, where the tin side of glass is generally slightly more reflective due to the diffusion of tin into that surface.

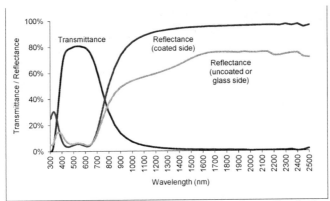

Figure II –Transmittance and Reflectance Spectra for one type of spectrally-selective glass -- a so-called high-transmission, low-emissivity silver-based coated glass

From the measured spectra, metrics that describe transmitted and reflected color can subsequently be calculated. In addition to the measured spectra, color metrics incorporate characteristics of the source of light and account for the typical physiology of human vision. These elements are important because changes in the nature of the light source and changes in the viewing perspective of the observer can influence color perception. What may not be accounted for is how, with changing angle of incidence and variable environmental conditions, the spectral energy distribution of sunlight may change during the course of a day. Color metrics do not account for personal preference of the observer, a product of the life experience of the observer, which also influences color perception.

A commonly used and universally accepted system for describing color is the L*a*b* color space, or CIELAB, developed by CIE, the International Commission on Illumination. The color space shown in Figure III is a three dimensional sphere comprised of three numerical coordinates:

- L* indicating lightness (+) and darkness (-).
- a* indicating red (+) and green (-).
- b* indicating yellow (+) and blue (-).

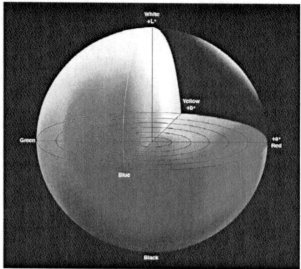

Figure III – L*a*b* Color Space, Konica Minolta Sensing, Inc., Precise Color Communication, 2007; graphic reproduced from Reference 1.

Descriptions of the calculations for L*a*b* values can be found in many textbooks and international standards. ASTM E308 is an example of one such standard.[2]

In addition to having a system for describing color, it is important to have a system for determining color acceptability. Color tolerance, or color difference, systems have been developed to compare a measured color to an expected, or target value. These systems can produce a single value of color difference that guides manufacturing in determining when product color is acceptable. The use of a color tolerance system also helps to produce a consistent product during the course of a production run, and each time the product is made.

The simplest form of a color tolerance system produces a color difference value, ΔE, that is the vector sum of the individual color differences between the sample color and desired values of L*, a*, and b* (Eq. 1). The appropriate value of ΔE may be identified by making visual comparisons, or may evolve based on acceptance of color difference in the marketplace.

$$\Delta E^*_{ab} = \sqrt{(\Delta L^*)^2 + (\Delta a^*)^2 + (\Delta b^*)^2} \quad \text{(Eq. 1)}$$

More complex systems, such as CMC, or CIE2000, incorporate additional factors to refine the basic CIELAB determination.[3]

The color tolerance space can be displayed graphically, and generally takes the shape of an ellipse in 2 dimensions (a*b*), or an ellipsoid in 3 dimensions (L*a*b*).

COLOR CONTROL IN FLOAT GLASS MANUFACTURING

Various colors, optical properties, and solar properties are achieved in float glass by the addition of metal oxides. Iron, cobalt, selenium, nickel, and chromium are but a few of many metal oxides that can impart color to glass. Figure IV illustrates the areas of color space that are influenced by some typical glass colorants.

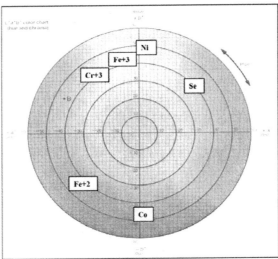

Figure IV Typical colorants in glass and the areas of color space that are influenced. The origin (0,0) represents a exactly neutral, colorless aesthetic.

The addition of each coloring oxide in the glass batch must be carefully metered, and contributions of coloring oxides from other sources, such as contaminants in raw materials, or in glass cullet returned to the melting process must be carefully accounted for. In addition to carefully controlling the colorant input, regular, diligent monitoring of the color and composition of the produced glass is also a key component of the color control scheme.

The amounts of each colorant in the glass can range from a few parts per million (ppm) to more than 1% by weight, and must be maintained in the right proportion. For some colorants, variation of as little as ±1 ppm can result in a noticeable change in the glass transmitted color. Additionally, effects from thermal variation in the melting process, combustion stoichiometry, and other chemical interactions, including the reduction-oxidation of glass must be considered as part of the color control process. It is therefore critical that methods are in place to precisely monitor and control glass colorant content.

One method for determining glass composition and glass colorant content is x-ray fluorescence spectroscopy (XRF). This method can be utilized to determine glass colorant content with a high degree of accuracy. However, this analysis method may not always be the most desirable in a production environment due to the cost of equipment, and the delay in acquisition of results that can accompany sample preparation and analysis.

Transmittance measurement of the glass can be accomplished quickly, with little sample preparation, and with minimal capital investment. Methods of interpreting transmittance measurements, and transmitted color, have been developed to estimate glass colorant content with good accuracy and minimal delay.

Each coloring oxide in glass has a unique wavelength-dependent absorption characteristic. The absorption coefficients for some typical glass colorants are shown in Figure V. The absorption curves indicate the wavelengths that are impacted by the colorant, as well as the intensity, or strength, of the colorant. In Figure V it can be seen that selenium (Se) is a strong colorant with a peak absorption near 500 nm, compared to cobalt (Co) with peak absorption between 600 and 650 nm, and ferrous iron (Fe^{+2}), with a broader absorption from the long wavelength visible range into the near infrared.

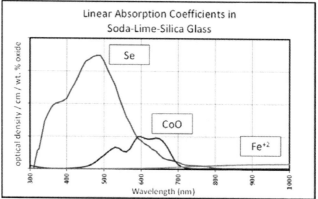

Figure V – Absorption characteristics of typical glass colorants

With knowledge of the absorption characteristic and content of a particular colorant in glass, the resultant transmittance of the glass can be estimated. Furthermore, the separate absorptions from multiple colorants in glass are additive, and the composite transmittance of glass with known amounts of multiple colorants can be determined. Figure VI illustrates the difference in transmittance spectra between clear glass with about 0.1 weight % iron, and bronze glass with higher iron content (about 0.3 wt. %), and small amounts of selenium (< 20 ppm) and cobalt (~ 30 ppm).

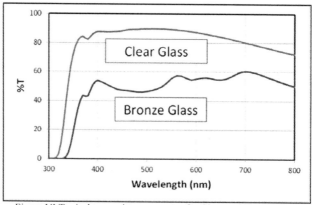

Figure VI Typical transmittance spectra for Clear and Bronze glass

From the composite transmittance spectra, the expected transmitted color can be calculated. Figure VII shows the distinctly different a*b* color coordinates for typical clear and bronze glass.

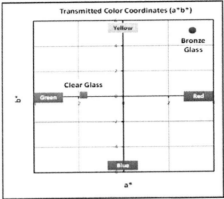

Figure VII Transmitted color coordinates in a*b* space for typical Clear and Bronze glass

By searching the literature, and evaluating color and composition of production glass and laboratory glass melts, a library of absorption data for glass colorants can be assembled. Using this library, models can be developed whereby glass color can be estimated for multitudes of coloring oxide contents and combinations, or conversely, coloring oxide content can be estimated based on evaluation of transmittance spectra. It is then possible to develop tools for use in glass production so that colorant content can be estimated from on-site measurement of transmitted color. This enables production personnel to make informed decisions regarding adjustments to the necessary colorant input.

One such tool is a color vector plot, shown in Figure VIII. A color vector plot shows the expected change in a*b*, (or L*), magnitude and direction corresponding to changes in glass colorant content. The vectors are calculated by modeling changes to each colorant.

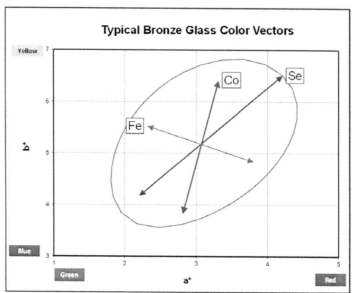

Figure VIII – Color vector plot for typical bronze glass

Measurements of the transmitted color of production glass can be plotted in L*a*b* color space, and the color coordinate position can be evaluated relative to the desired coordinates by applying vector analysis. The change in colorant content that is necessary to produce the target color can then be determined iteratively by manual or algorithmic routines. This control scheme, combined with an appropriate color tolerance system, assures that consistent glass transmitted color is produced. This translates to acceptable building aesthetics when uncoated glass is used in a window or curtainwall. Consistent substrate glass transmitted color will also aid in reducing reflected color variation in the manufacture of coated glass.

COLOR CONTROL OF MULTILAYER THIN FILM COATINGS ON GLASS

Adding one or more layers of thin films to glass increases the complexity of the interaction of light with glass. A film is considered "optically thin" if its optical thickness (the product of the film's geometric thickness and its wavelength-dependent refractive index), is less than or on the order of the wavelength(s) of light of interest. In contrast, a film is considered "optically thick" if its optical thickness is larger than the so-called coherence length of the incident light.

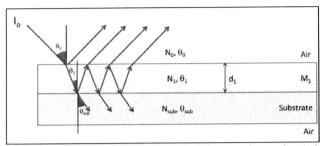

Figure IX – Interactions between incident light and a single layer coating on glass (original graphic courtesy of A.V. Wagner, PPG Industries, Inc.)

Considering the simplest example of a single thin film layer on glass (Figure IX) the refractive index difference on either side of each interface causes a portion of light to be reflected. The refractive index discontinuity determines the amplitude and relative phase of the reflected and transmitted light waves at each interface. The amplitudes and relative phases associated with reflectance from, and transmission through, each interface are determined by the interface's so-called Fresnel coefficients, which are derived from Maxwell's Equations of electromagnetism. Note that the Fresnel coefficients are also dependent upon the polarization of the light and the angle of light incidence at each interface.[4]

The additional phase lag accumulated by a light wave which traverses the film is described by the layer's phase thickness parameter, δ_j:

$$\delta_j = \frac{2\pi \tilde{N}_j d_j \cos \theta_j}{\lambda_o} \text{ (Eq. 2)}$$

where the material-dependent $\tilde{N}_j$, d_j, *and* θ_j are the complex refractive index, the geometric thickness, and the angle of propagation of the light in layer j, respectively, and λ_o is the wavelength of the light in vacuum. The complex refractive index $\tilde{N}_j$ of each layer comprises a real and imaginary part:

$$\tilde{N}_j = n_j - ik_j = \text{complex refractive index of layer } j \text{(Eq. 3)}$$

Where n_j = refractive index of layer j, and k_j = extinction coefficient of layer j.

The real part of the complex refractive index, n_j, is loosely referred to as the material's refractive index. The imaginary part of the complex index, k_j, referred to as the material's extinction coefficient, is a parameter which characterizes the absorption of light in the material. Both n_j and k_j, are, in general, functions of the light's wavelength. In addition to amplitude and phase changes which occur at interfaces between coating layers, and the accumulated relative phases associated with traversal of the light waves through the coating layers due to each layer's phase thickness, it is essential to also account for the attenuation of the light waves when any given layer exhibits non-zero extinction coefficient.

In view of the above considerations, the interaction of light with thin film coatings will, in general, be a function of incident angle, polarization, and wavelength of light. The reflections from the various interfaces within the coating interfere coherently with each other to produce the net reflectance and transmitted spectra. With knowledge of the thickness of the layer and the wavelength-dependent optical properties of the material in the single layer coating, along with similar information about the substrate glass, it is possible to express the reflectance and transmittance as a closed-form expression[3], from which the reflected and transmitted color coordinates can be calculated.

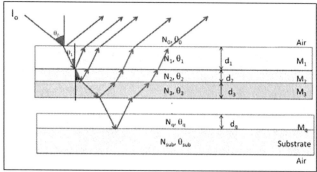

Figure X – Interactions between incident light and multilayer coatings on glass

Multilayer coatings on glass present additional challenges for producing consistent and predictable color due to a multitude of optical interactions between layers, as illustrated in Figure X. In addition to the factors previously mentioned for a single layer coating, the sequence of the various materials in the coating structure also affects the color of the coated substrate. For coatings comprising multiple thin film layers, a well-developed transfer matrix formalism is used wherein the net reflectance and transmittance are calculated from the product of 2 x 2 matrices acting on a 1 x 2 column vector which represents the substrate:

$$\begin{bmatrix} B \\ C \end{bmatrix} = \left\{ \Pi_{j=1}^{q} \begin{bmatrix} \cos \delta_j & \dfrac{i \sin \delta_j}{\eta_j} \\ i \eta_j \sin \delta_j & \cos \delta_j \end{bmatrix} \right\} \begin{bmatrix} 1 \\ \eta_{sub} \end{bmatrix} \text{ (Eq. 4)}$$

Each thin film layer is described by its own 2 x 2 matrix. In addition to the phase thickness, δ_j, each film's matrix includes a parameter η_j known as the film's optical admittance, which is a function of the film's complex refractive index, $\tilde{N}_j$, and the incident angle and polarization state of the light. The substrate's 1 x 2 column vector contains the optical admittance of the substrate, η_{sub}. The substrate's 1 x 2 column vector does not reference a phase thickness because the substrate is "optically thick" and it is not meaningful to speak of the relative phases of light waves traversing such an optically thick layer. The net reflectance and transmittance, resulting from the optically-coherent light waves reflected and transmitted from the thin film structure, are calculated using the B and C components of the 1 x 2 product vector. In contrast, reflectance and transmittance from the interfaces of optically thick layers, such as the macroscopically thick substrate, are calculated by incoherent addition of light intensities, not coherent amplitudes as is done for the thin film structure. The details are beyond the scope of this paper; we refer the reader to Reference 6. The matrix calculation must be repeated at each wavelength, angle, and light polarization of interest. Closed-form analytic expressions for reflectance and transmittance do not exist for multilayer coatings. Therefore, computer-based numerical models are necessary. A number of commercially-available software packages, which run on standard personal computers, exist for this purpose. Such computer-based optical models can be employed for process control in manufacturing operations to ensure all coating layers stay on-target vis-à-vis thickness and lateral uniformity.

In light of the preceding discussion, the reflectance and transmittance spectra, and hence reflected and transmitted color, of a coated glass depends upon the: (1) geometric thickness of each layer, (2) lateral thickness uniformity of each layer over the surface of the substrate, (3) optical properties (i.e. refractive index and extinction coefficient) of each layer, and the lateral uniformity of those properties, (4) spectral distribution (relative power versus wavelength) of incident light, (5) angle of incidence of light, or equivalently the angle at which the observer views the coated substrate, and (6) the polarization state of the incident light.

It is essential to tightly control the thickness uniformity of each layer of the coating over the surface of the substrate. Flat glass for architectural glazing and other applications (e.g. automotive, solar) is typically produced and sold in lateral dimensions up to about 100 inch x 144 inch. One illustration of the effect of layer thickness/uniformity is shown in Figure XI which shows the glass-side reflected a*b* chromaticity plot, generated using an optical computer model, for a thin film-coated glass at near-normal angle of incidence. When produced on-target, the coated substrate has a glass-side reflected color in the center of the red ellipse. The solid black curve traces out a trajectory which shows how color moves in the a*b* plane as the thickness of one of the coating's layers is increased (solid circles) or decreased (open circles) in 2% steps. In this example, the glass-side reflected color can tolerate up to about an 8% deviation in the thickness of this particular layer while remaining within the desired manufacturing color tolerance, assuming no other layers are off-target. However, for certain multilayer thin film coatings, the thickness of some layers must be controlled at tolerances of ≤ 2%. Proprietary technologies have been developed to achieve this level of lateral uniformity and process stability. In the case of, for example, a 10 nm thick layer, this requires a thickness uniformity of about ± 0.2 nm, which is about the diameter of individual atoms, over the width of the coated glass substrate.

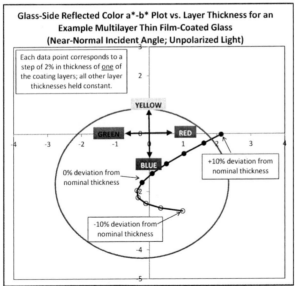

Figure XI – Effect of layer thickness variation on reflected color at fixed angle (near-normal incidence) for a thin film coated glass; unpolarized incident light.

The color/aesthetics of thin film-coated glass products are a function of angle-of-incidence, the viewing angle of the observer, or a combination of the two. As an example, Figure XII shows the color trajectory, generated via an optical model, for a coated glass as a function of incident angle. The black curve shows how the glass-side reflected color moves in a*b* space in 5° steps of incident angle. At normal/near-normal incidence, the reflected a*b* color coordinates are (a*,b*) = (-0.2, -1.7). As the incident angle moves from 0° to 85°, the glass-side reflected color moves as shown. This variation in color with angle is not symptomatic of a "defective" coating; rather, it is an intrinsic characteristic of the optical interference physics. Accordingly, the coating design must meet the aesthetics requirements at oblique angles in addition to at near-normal incidence.

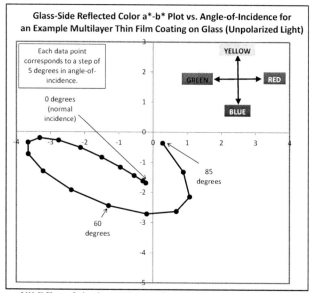

Figure XII Effect of viewing angle on reflected color of a thin film-coated glass

For architectural structures employing large expanses of glass such as all-glass facades, it is important to ensure that adjacent portions of such glazings exhibit identical variation in color as a function of incident or viewing angles. Thus, it is critical to ensure: (1) that the coatings on adjacent portions of the glazings are all "on-target", and (2) sufficient lateral uniformity of the thicknesses and optical properties of the coating's layers, such that the color of adjacent areas of the coated glazing all move identically as a function of incident angle. Indeed, as an example of an undesirable occurrence, it is possible to produce coatings having identical layer stack sequences, but distinctly different layer thicknesses, such that the aesthetics of those coatings may match at one or more incident angles (e.g. near-normal incidence), but exhibit different color trajectories, and hence color-mismatches, at other angles of incidence. In addition, such "degenerate" coatings can have disparate color trajectories when the thickness of any particular coating layer is varied. Therefore, as more sophisticated coatings with greater numbers of layers have become more commonplace, proprietary process control methodologies have been developed to discriminate between such degenerate coatings and uniquely identify which coating is being produced, thereby enabling the production coater operator to confidently control the process.

The preceding discussion has focused largely on the spectral properties of coated glass and the variables that affect their reflectance and transmittance spectra, and hence color. With regard to the characteristics of the incident light, in addition to the wavelength parameter, the polarization state of the incident light can affect the perceived and measured color of an object.

Light, an electromagnetic wave, comprises a sinusoidally oscillating electric field which is perpendicular to the wave's direction of propagation while the companion magnetic field oscillates perpendicular to the electric field. At oblique angles of incidence, a plane, known as the plane-of-incidence, is defined by the incident propagation vector, the reflected propagation vector, and the surface normal. Light having its electric field oriented perpendicular to the plane-of-incidence is referred to as s-polarized (or TE = transverse electric) light; p-polarized (or TM = transverse magnetic) light corresponds to the case in which the electric field lies in the plane-of-incidence. Unpolarized light can be mathematically represented as an equal mixture of s-polarized and p-polarized light. Daylight, which is usually largely unpolarized, under certain atmospheric and solar position conditions, may exhibit measurable polarization, can have consequences for the perceived color of thin film-coated glass products as well.

The effect of light polarization on the color of thin film-coated glasses can be illustrated by the example shown in Figure XIII. The black curve/circles show the glass-side reflected color trajectory as a function of angle of incidence for the case of unpolarized incident light, modeled as an equal mixture of s-polarized and p-polarized light; those data points are identical to the data plotted in Figure XII. The blue curve/circles shows the color trajectory for the case of 100% s-polarized incident light. The pink curve/circles are for the case of 100% p-polarized light. All three color trajectories converge as the angle of incidence approaches zero; there is no distinction between s-polarization and p-polarization at normal incidence. Thus, under certain conditions, the perceived color of a coated glass can depend upon the polarization of incident daylight and the viewing angle.

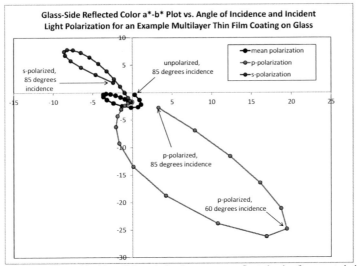

Figure XIII – Effect of incident light polarization state on reflected color for a coated glass

SUMMARY

Requirements for achieving the desired aesthetics for glass in buildings include producing the color that is expected by the customer, assuring that the color is uniform within and between each unit of window glass, and assuring that the change in color of the window glass with multiple viewing angles is also acceptable.

Prevention of non-uniform and undesirable building aesthetics can be accomplished by carefully controlling the input of glass colorants when manufacturing float glass, and precisely controlling layer thicknesses in the manufacture of multi-layer coatings on glass. Both of these operations require diligent monitoring and evaluation of transmitted and reflected color, and the employment of control schemes to maintain color within limits of acceptability.

Sophisticated tools and standards for color design and color control for float glass and multi-layer coated glass have been developed and are in use.

REFERENCES

1. Precise Color Communication, Konica Minolta Sensing, Inc., 2007
2. ASTM E308 – 08, "Standard Practice for Computing the Colors of Objects by Using the CIE System"
3. CIE Draft Standard DS 014-6/E:2012, Colorimetry - Part 6: CIEDE2000 Colour-Difference Formula
4. E. Hecht, *Optics*, 2nd ed., Addison-Wesley Publishing Company, 1987.
5. O.S. Heavens, *Optical Properties of Thin Sold Films*, Dover Publications, 1991.
6. A. Macleod, *Thin Film Optical Filters*, 2nd ed., McGraw-Hill/Adam Hilger Publishing, 1989.

HARD GLASS – THERMAL STRENGTHENING OF CONTAINER GLASS

Steven Brown
Emhart Glass Research Center
123 Great Pond Drive
Windsor, CT USA

ABSTRACT

Glass is the premier choice for beverage and food packaging. It offers elegance, beauty, complete recyclablity, long shelf life and inertness to the food or beverage that it contains. Unfortunately, it can break into sharp fragments, sometimes with seemingly minimal provocation, possibly leading to consumer injury or filling-line down time.

The strength of pristine glass is actually very high measuring upwards of 2,000,000 psi. However, its practical strength is limited by the severity of its defects, and, can fracture in the presence of tension. It is expensive and difficult, if not impossible with today's technology to produce defect-free container glass so the effective glass strength is usually more on the order of 2,000 to 25,000 psi depending on the extent and location of the defects.

The glass industry has been loosing market share to plastics, aluminum cans, and paper cartons, especially in the US as a result of breakage hazards and the required container weight to minimize breakage.

Emhart Glass has spent the past decade engineering a process that strengthens annealed containers by first heating the articles to above the annealing point (usually between 530°C – 550°C for soda lime glass) followed by rapid cooling to below the glass strain point (usually between 500°C and 520°C). This results in a residual compression-tension-compression parabolic stress sandwich, similar to tempered flat glass that increases the breaking strength commensurate with the level of imparted surface compression levels.

INTRODUCTION

Tempering of glass has been well established for decades but very little work has been done in the area of container glass. This is due to the difficulties associated with rapidly cooling the inside surfaces of the container through the limited opening of the finish, the wall thicknesses and variations thereof, and due to the complex geometries involved.

There are actually four groups of tempering stress levels in the glass industry as defined by the American Society for Testing & Materials (ASTM).

a. Annealed Glass – Typically has close to zero stress but practically speaking is usually within +/- 5MPa of zero (+/-725 psi). This is the standard for all container glass in use today. It produces a stable container but requires fairly substantial wall thicknesses, resulting in increased weight, transportation costs, and a larger carbon footprint. Upon fracture, annealed glass typically fails with large shard-like pieces of glass.

b. Heat Strengthened Glass – Typically has surface compression levels between 24MPa (3.5 ksi) and 69MPa (10 ksi). Heat strengthened glass will still break in a shard-like fashion but the pieces will be smaller than a comparable annealed fracture. The shape of the stress field through the thickness of the glass is parabolic having a buried peak tensile stress on the order of 12MPa (1.7 ksi) to 35 MPa (5 ksi).

c. Tempered Glass – Typically used in the automotive and architectural industries where strength is important as well as a "dicing" fracture pattern upon failure. Tempered glass is classified as having a surface compression stress range between 69MPa (10 ksi) and 103MPa (15 ksi). The shape of the stress field through the thickness of the glass is

119

parabolic in shape having a buried peak tensile stress on the order of 35 MPa (5 ksi) to 52 MPa (7.5 ksi).

d. Safety Glass – Characterized as having surface compression levels greater than 103 MPa (15 ksi). Safety glass is typically used where the glass thickness can be thicker than ¼". It has the highest strength of all the tempering categories listed above.

Fragment Size:
Annealed > Heat Strengthened > Tempered > Safety Glass

Strength:
Safety Glass > Tempered > Heat Strengthened > Annealed

As a general rule of thumb, heat strengthened glass has twice the strength of annealed glass (for flat plate) and tempered glass has four to five times the strength of annealed glass (for flat plate).

The Parabolic Stress Profile
The parabolic shape of the stress distribution through the wall for either heat strengthened, tempered, or safety glass is governed by the following equation:

$$\sigma = \sigma_m [1-12(1/4-t/h+(t/h)^2)]^{(1)} \qquad [1]$$

Where:
σ = Stress at the point of interest
σ_m = Stress at the mid plane
t = thickness at the point of interest
h = wall thickness

Therefore, at t=h/2 (the mid-plane), $\sigma = \sigma_m$
And at t=0 or t=h (either surface), $\sigma = -2 * \sigma_m$

From this equation, it can also be shown that the thickness of the compression layer is theoretically 21% of the overall wall thickness. Therefore, 42% of the glass will be in compression and 58% will be in compensating tension. The net stress has to be zero so the beneficial surface compressive stress is balanced by the buried tensile stress. Also, due to the cylindrical symmetry of most glass containers, and their small wall thickness in relation to their diameter, the stress will be governed by the following summation rule:

$$\sigma_{axial} = \sigma_{radial} + \sigma_{circumferential} \qquad [2]$$

Where: σ_{radial} = near zero and is usually negligible

A typical parabolic stress curve is shown in figure 1 below.

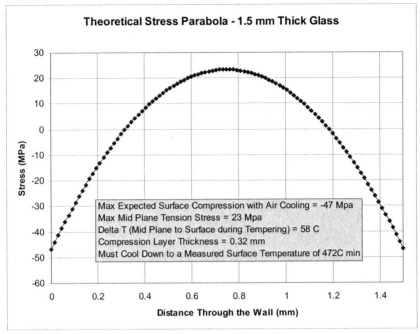

Figure 1 - Theoretical Stress Curve

What's important to realize is that if the cooling on the outside surface of a container is greater than the cooling on the inside surface, then the resulting stress parabola will favor the outside surface. The net stress will still be zero but in extreme unbalanced cooling situations, the thickness of the compression layer on the inside surface can be dangerously thin or perhaps even non-existent. In this worse case scenario, the "strengthened" container could actually be weaker than an annealed container depending on where the critical defects reside in relation to the residual tensile stresses present at the surface. But the goal is to achieve enough compression layer thickness on either surface to contain the typical distribution of container defects found in the industry. It will therefore take more applied external stress to overcome the residual compression as compared to an annealed glass under the same conditions. In general, the mechanical strength of glass is inversely proportional to the square root of the flaw length.

Next, let's consider a typical glass bottle having a minimum wall thickness of 1.5 mm. If cooled, quickly and evenly, the imparted surface compressive stress can typically be in the range of 25 MPa with a buried tensile stress of 12.5 MPa. This would require a heat transfer coefficient achievable with forced air cooling. The thickness of the resulting compression layer will be 0.21 x 1.5 = 0.315 mm which is thick enough to cover all of the typical container defects as shown in the following table.[2]

Table I - Typical Flaw Depths			
Condition	Microns	mm	inches
As Formed	10	0.01	0.00039
Shipped	55	0.055	0.00217
Used Once	80	0.08	0.00315
Returnable (Visible Damage)	105	0.105	0.00413

In this instance, the minimum compression layer thickness is at least three times the depth of typically found defect sizes in the returnable bottle market so in theory the defects should be surrounded by compressive stresses and will therefore have higher strength than comparable annealed ware.

The key to establishing an adequate level of surface compression is therefore to start at a high enough temperature, well above the annealing point, then to rapidly and evenly as possible cool all surfaces, inside and out, at a constant rate until the entire bulk of glass is below the strain point. As the glass cools through the transformation range, the temperature difference between the surfaces and the core is critical. In fact, the resultant residual surface compression can be estimated for soda lime glass using this simple equation:

$$\sigma = 0.8 * \Delta T \qquad [3]$$

Where:
σ = Surface Stress in MPa
ΔT = Temperature difference, surface to core, degrees C

Measuring the Stress Profile

Emhart Glass uses an Immersion Polariscope, built by GlasStress Inc. located in Tallinn Estonia, to completely map and characterize a treated bottle. The device is shown in figure 2 and a sample plot is shown in figure 3.

Figure 2 - GlasStress AP-07 Immersion Polariscope[3]

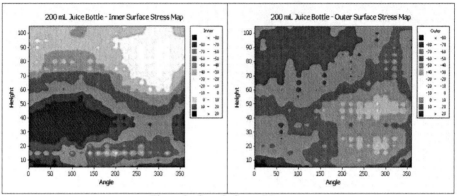

Figure 3 - Sample Stress Map

In figure 3, it is apparent that the stress is not uniform but tends to follow the wall thickness (especially on the outside surface), is affected by changes in geometry and the cooling hole pattern used in the quenching process. The overall average level of stress on both surfaces is about 50 MPa. This is sufficient to increase typical bottle strengths as compared to comparable annealed ware as follows:

Impact:	5% - 30%
Internal Pressure:	20% - 60%
Vertical Load:	20% - 100%
Thermal Shock:	>50%
Drop:	>100%; Significant improvement in terms of the number of survivors from a given drop height

In all situations, the upper strength limits are achieved in heavier glass articles having glass thicknesses averaging 3.2 mm (1/8 inch) while the lower strength limits are seen in lighter weight articles having wall thicknesses approaching 1.2 mm (0.050 inches).

The Effects of Shape & Size on Performance

Emhart Glass has successfully strengthened soda lime containers ranging in volume from 100mL to 750mL, with throat openings between 16mm to 70 mm and in the following colors: flint, various shades of green, amber, yellow and blue.

The geometries that tend to limit the feasibility of the strengthening process are deep decorative engraved features, wall thickness less than 1.2 mm, large thickness variations within a given container, non-rounds, sharp corners, and small throat openings compared to the bottle volume or diameter. The most difficult of these parameters to overcome is the throat opening since this limits the amount of cooling air that can both enter and exit through the throat to cool the inside surfaces of the container.

In addition, the container must be a high quality to endure the strengthening process since it will undergo a large temperature gradient during the strengthening process where temporary tensile stresses can run quite high. Containers that already have sharp-tipped checks or cracks will sometimes fail during the cooling process.

The effect of wall thickness on the resulting residual stress is significant as thinner sections require more cooling than thicker sections. The relationship requires that the cooling rate vary by the square of the glass thickness. For example, the cooling rate would have to increase by a factor of 9 if the glass thickness varies from 3mm to 1mm.

Examples
    A. 100mL Cordial
    In this series of tests, a decorative 100mL cordial beverage bottle weighing only 80 grams was heat strengthened and the burst results were improved by 45% as shown in figure 4.

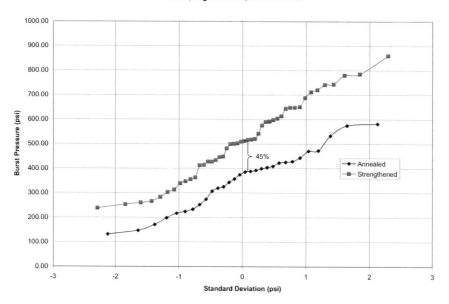

Figure 4 - 100mL Cordial Burst Pressure Curve

    B. 200mL Juice
    A 200mL contoured-shape juice bottle weighing 130 grams was strengthened and compared to annealed bottles for burst, drop and impact loads. The results are shown in figures 5-7.

200mL Juice, 130g, Burst Results

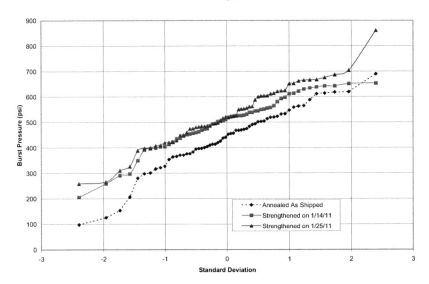

Figure 5 - 200mL Juice Burst Pressure Curve

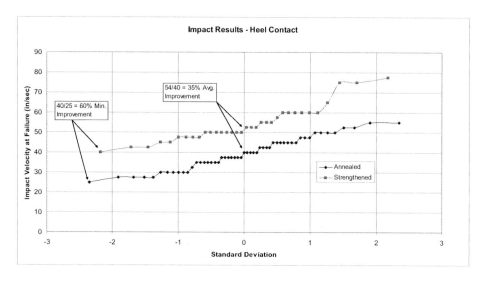

Figure 6 - 200mL Juice - Impact Results

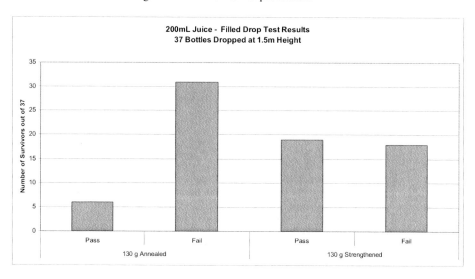

Figure 7 - 200mL Juice - Drop Test Results

C. 330mL Beer

A commercial 245gm, 330mL beer bottle was weight reduced to 209 grams and strengthened. The burst, impact and drop results are shown in figures 8-10.

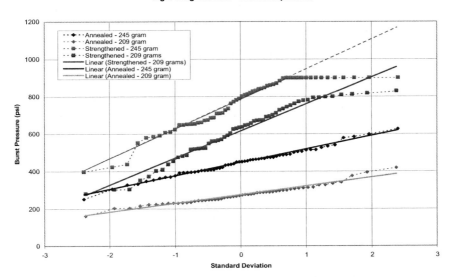

Figure 8 - 330mL Beer Burst Comparisons

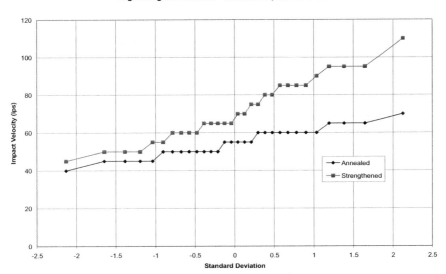

Figure 9 - Light Weight 330mL Beer Impact Results

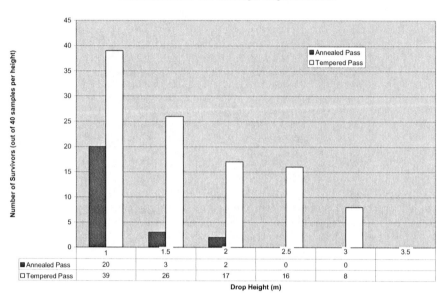

Figure 10 - Light Weight 330mL Beer Drop Test Results

D. 600mL Returnable Beer

A returnable 600mL, 440 gram beer bottle was strengthened and burst tested. The results are shown in figure 11. Limited shoulder impact tests were also conducted where the average impact strength was increased by 15%.

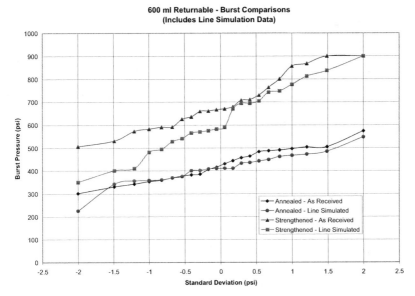

Figure 11 - 600mL Returnable Beer Bottle Burst Data

Process Definition

In the heat strengthening process annealed bottles are heated to above 600°C as quickly as possible to avoid deformation then rapidly cooled to less than 500°C. During this cooling process, the interior and exterior surfaces of the glass are cooled quicker than the glass that is buried between the surfaces. This results in a parabolic temperature distribution through the glass which gets locked in as a stress distribution once the temperatures are dropped to below the strain point (usually between 500°C-520°C for soda lime) at the surface.

Cooling is achieved by compressed air traveling through specially designed nozzles and orifices to cool the outside surface, inside surface and the bottom of the bottle. The cooling times vary depending on bottle geometry and wall thickness but typically range from 9 seconds to 15 seconds for bottles having a capacity range of 100mL to 1 liter.

CONCLUSIONS

Emhart's strengthening process will work on the majority of the beverage and food containers in production today with savings to be realized in terms of reducing weight without compromising strength, life, aesthetics, or performance. The first commercial implementation will be a secondary process capable of producing approximately one million containers per month. The next generation machine will be in-line with the bottle production process and will match typical current production speeds.

REFERENCES:

1. Aben, H., "Photoelasticity of Glass" Lecture Notes, June 11-13, 2003, Laboratory of Photoelasticty, Institute of Cybernetics, TTU, Tallinn, Estonia, page 206, equation 12.13.
2. Kurkjian, C., "The Glass Researcher", Alfred University, Spring 2002
3. Aben, H., Anton J., Errapart, A., GlasStress Ltd., Tallinn, Estonia

EFFECTS OF GLASS COATINGS ON THE GLASS TEMPERING PROCESS

Charles Cocagne
Recent Additions, Inc.
Ann Arbor, MI, USA

ABSTRACT

In recent years flat glass manufacturers have developed numerous glass coatings for energy savings, architectural enhancements, scratch resistance, water shedding, anti-reflectivity and other applications. These value-added coatings can be either hard coats applied in the glass manufacturing process or soft coats applied by vacuum deposition after the glass has been processed. These coated glass products have become a major part of flat glass sales in both the developed and the developing countries.

Coated glass products have opened new markets, but they also have had a major impact on the handling of the raw glass in preparation for tempering, laminating and fabrication of insulated glass products. The primary emphasis of this paper will be to combine a discussion on the glass tempering or heat treating process, glass heating techniques with respect to the basic glass properties, and the evolution of glass tempering systems to handle efficiently the new coated glass products.

HISTORY

The method of strengthening glass via thermal heat treating has been known for since the early 1900's. However, it took the demand for specific needs and applications before the development of commercial glass tempering began.

In the early 1950's television became an enormous part of our lives. However, early TV tubes had a tendency to explode. Tempered glass offered a solution that allowed passing the Underwriter Laboratories (UL) safety requirements. This product became the accepted safety standard for TV implosion plates.

During the late 1950's the automotive industry was advancing toward curved sidelites. This process had never been utilized before and was not covered by the American National Standard Z26.1, "Safety Code for Safety Glazing Motor Vehicles Operating on Land Highways." It took from 1957 to 1960 before the legislatures of all 50 states agreed that tempered safety glass under the Z26 standard could be used in car sidelites.

In the residential and commercial building sector more and more glass was being used in storm doors, combination doors, entrance-exit doors, sliding patio doors, closet doors, shower and tub doors and enclosures. Since there was no requirement for safety glass, reports of serious injuries were frequent, and in many cases legal action was taken. As a consequence, on July 6, 1977, the Consumer Products Safety Commission published CPSC 16 CFR 1201 – Safety Standard for Architectural Glazing Materials - to reduce or eliminate unreasonable risks of death or serious injury to consumers if glazing material were broken by human contact. Under the CPSC specification the glass fabricators were required to label all safety glass and certify the product to meet the requirements of 16 CF 1201.

THERMAL HEAT TREATING PROCESS

In order to fabricate tempered or heat-strengthened products, the raw glass is cut to the proper size; the edges are seamed or polished; holes, notches, etching, ceramic paints are added as required; and the product is marked to show compliance with applicable standards. After pre-processing is complete, all sides and edges of the glass must be washed prior to entering the glass tempering furnace. In the case of coated/insulated glass products, edge deletion of the coating is necessary before washing. Upon entering the furnace the glass is heated to near its softening point (620C or 1,150F) before being removed quickly and quenched with high pressure air. The primary purpose of this process is to make the glass stronger and more resistant to mechanical and thermal stresses and to make a safety glazing product. A schematic of the heat treating process is given in Figure 1.

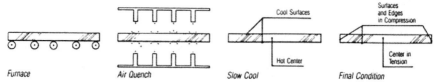

Figure 1: Heat Treating Process

   In the ideal world, if a properly supported sheet of glass could be heated to a uniform temperature and then quenched uniformly, it would exhibit a uniform degree of temper and be perfectly flat. However, the variables involved in the conveying, heating and quenching process are numerous; and obtaining a flat piece of tempered glass is a challenge which is more difficult on coated glass products.

   In most cases fully-tempered glass is required to meet the safety standards of CPSC 16 CFR 1201. For flat or bent tempered glass, fully tempered is defined as being heat treated to have a minimum surface compression of 10,000 psi (69 MPa) with an edge compression of not less than 9,700 psi (67 MPa). At this compression stress level the glass is approximately five (5) times the strength of annealed glass. The typical stress profile and break pattern are illustrated in Figure 2.

Figure 2: Fully Tempered Glass Stress Profile and Break Pattern

Since the 1990's there has been a trend toward the use of heat strengthened glass on commercial buildings, which is done for the purpose of retaining the glass in the insulated glass units when the glass fails. By definition, heat strengthened glass should have a surface compression between 3,500psi (24 MPA) and 7,500psi (52 MPa). At this stress level the glass is approximately three (3) times the strength of the raw float glass. The typical stress profile and break pattern are illustrated in Figure 3.

Figure 3: Heat Strengthened Stress Profile and Break Pattern

GLASS – THE BASIC RAW MATERIAL

G.W. Morey, an early 20th century glass technologist, defined glass as "…an inorganic substance in a condition which is continuous with and analogous to the liquid state of that substance, but which as a result of a reversible change in the viscosity during cooling has attained so high a degree of viscosity as to be for all intents and purposes, rigid." ASTM defines glass as "…an inorganic product of fusion that has cooled to a rigid condition with crystallizing."

In either case, float glass, like all other glasses, is defined as a liquid and has specific viscosity versus temperature curve. Most of the flat glass products are of a soda lime composition and are defined as shown below in Figure 4.

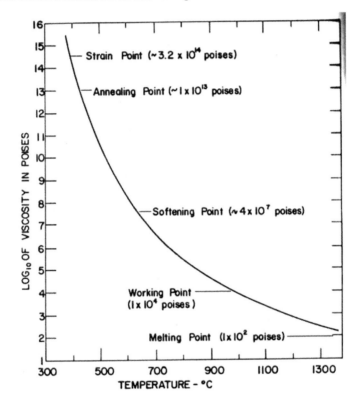

Figure 4: Typical Viscosity Curve for Glass

Table 1: Typical Characteristics for Soda Lime Glass

Characteristics	Temperature	Viscosity (Log10-Poises)	Description
Melting Point	1,530°C (2,800°F)	2	Point at which raw materials are melted to form glass
Working Range	710° - 1,090°C (1,300° - 2,000°F)	4 – 7.6	The range between the working point (2,000°F) and the softening point (1,300°F) in which the glass can be formed into a commercial product
Tempering Pre-Heat	620°C (1,150°F)	7	Point at which the glass can be quickly cooled to induce compressive stresses into the glass
Annealing Point	550°C (1,020°F)	13	The point at which glass annealing can begin
Strain Point	510°C (950°F)	14.5	The point at which permanent stress is fixed in the glass product

Optical properties of glass include reflectance, absorption and transmission. The property of reflectance is where optical coatings can have a pronounced affect on the heating glass during the glass tempering process. In the past, due to the high infrared absorption characteristics of soda lime glass, most glass tempering furnaces used infrared radiation from heating elements to heat the glass. However, many of the new glass coatings are designed to reflect a large portion of the infrared spectrum. This reflectance has made it very difficult (and in some cases almost impossible) to process the new low emissivity glasses in conventional furnaces.

Thermal expansion is another property which becomes very important during heating in the glass tempering process. For example, a 72" long glass will expand approximately 3/8" while being heated to glass tempering temperature. A differential in temperature in the glass thickness or glass length can cause the glass to bow up or down and create problems during the quenching process.

Thermal conductivity of the glass is the limiting factor when heating by convection or conduction. Due to glass heating problems encountered on low emissivity and other coated glasses, there has been a movement toward convection heating. Theoretically, when using convection heating, clear glass and coated glass products will heat at the same rate; however, a true 100% convection furnace is impossible construct.

GLASS TEMPERING FURNACES

After the development of high-grade fused silica rolls, most glass tempering furnaces have evolved to a roller hearth design. These furnaces have been designed to convey the glass on horizontal rolls into the furnace and into the quench section after heating to the proper temperature. A discussion of the components and some of the associated problems are given below.

Conveyance

At present, the primary means of conveying glass in the glass tempering or heat treating operation is on horizontal fused silica rolls. Fused silica has become the material of choice because it has very low expansion in the glass tempering range. Even though these rolls can be manufactured with a very low out of tolerance (.002 inch or lower), these rolls still can be the source of roller wave distortion. This type of distortion has a pitch equal to the roller circumference as shown in Figure 5. In addition, if the rolls are not properly maintained, they can leave marks on the glass and can cause a difference in strength between the upper and lower surface of the glass and a possible source of bowing or warping when the glass is quenched.

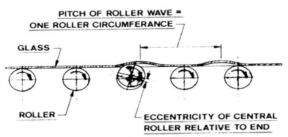

Figure 5: Roller Wave Distortion

In the past some furnaces were produced in which the glass was floated on a film of hot air or products of combustion as it passed through the furnace. In these furnaces the parts were driven by tilting the support bed laterally with respect to the direction of flow so that the lower edge of the part tends to rest against the conveying means. On these furnaces the support film was approximately 0.010 inches thick; consequently, great care was needed to assure that the parts were heated equally and uniformly on both sides to prevent scraping of the part as it passed through the furnace. These furnaces produced a high quality product but were not very flexible when running different parts, shapes and glass thicknesses.

Heating and heating effects

The primary method of heating the glass in the tempering process has been by radiation supplied by electrical-powered resistance elements or gas burner systems. However, the flat glass manufacturing sector has developed numerous glass coatings for high performance glass applications. These value-added glass coatings can be either hard coats applied in the glass in the manufacturing process or soft coats applied by vacuum deposition. In many cases these coatings provide a high reflectivity to infrared radiation. Therefore, in many cases the traditional glass tempering furnace cannot effectively heat treat these products. These new coatings have caused an evolution from the typical radiant furnace toward a true convection heat furnace. A summary of this evolution is given in Table 2.

Table 2: Glass Tempering Furnace Comparisons

System	Radiant Furnace	Radiant with Convection (cold air)	Radiant with Convection (hot air)	True Convection
Construction	Electric heating elements	Electric heating element + compressed air pipes blowing cold or preheated air into the furnace; moving air adds convective heating	Electric heating elements + piping for blowing hot turbocharged air into the furnace; moving air adds convective heating	Encapsulated heating elements + blowers circulating hot air in the furnace
Heating source	Direct radiation Natural convection Conduction	Direct radiation Improved convection Conduction	Direct radiation Forced convection Conduction	Indirect radiation Forced convection Conduction
Schematic Presentation				

As shown in Table 2 the furnace operator must contend with several heat sources while trying to balance the heat to the upper and lower surfaces of the glass. These heating sources include direct radiation, indirect radiation, conduction from the hot silica rolls and convection. In general, as the convection component increases, the glass heating rate increases, especially on coated glass products.

Quenching effects
As stated earlier, if a flat glass part could be heated to a uniform temperature and then quenched uniformly, it would stay flat and exhibit a uniform temper. This standard is nearly impossible, but in true convection furnaces some of the problems are minimized.

Examples of quench effects are as follows:
Hot edges – If a part has hot edges upon entering the quench, the edges will contract more than the center material and cause an oil-can effect.
Hot surface – If the glass has one surface hotter than the other, the quench will cause the glass to bow.
Hot middle – A part having a hot middle relative to its edges becomes saddle-shaped after it reaches room temperature.

COMPARATIVE FURNACE PERFORMANCE

The development of the high performance Low-E coatings has had a major impact on the glass tempering process. Before convection heating, it became almost impossible to heat the coated glass to the proper temperature. As a consequence there was slow development of adding convection to the radiant furnaces by means of heated compressed air aspiration systems. The Glaverbel heat balance system (GHBS) shown in Figure 6 is a typical system that has been added successfully to radiant furnaces to assist in the processing of soft coat Low-E products.

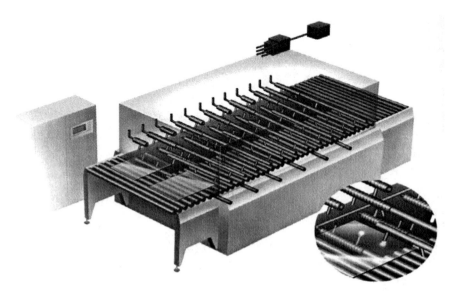

Figure 6: Radiant Furnace with GHBS System

With the need for higher production rates and improved product quality more and more forced convection was added to the basic radiant furnace. These modifications have been quite successful, but an experienced operator is needed to balance the heat forces.

Several full convection glass tempering furnaces have been developed. These furnaces are designed to have no direct radiation to the glass. These furnaces will typically have full convection above and below the glass, utilizing hot fans for recirculation of the hot air in the furnace. The best systems are designed to handle a high volume of heated air at a low velocity. At present these systems are limited to approximately 75% convection efficiency and are about 92% energy efficient. An example of an electric full-convection furnace is given in Figure 7.

Main components in the furnace

- hot air blowers (1)
- heating elements (2)
- convection air channels (3)
- nozzles (4)
- rollers (5)

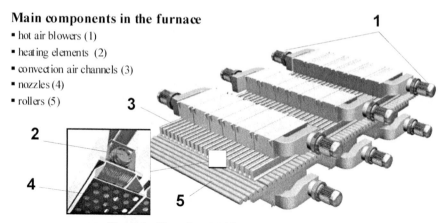

Figure 7: Glassrobots Multiconvection Furnace

Production rates on different types of furnaces are given in Figure 8 for both clear and soft-coated Low-E glasses. This data shows an evolution toward more efficient operation as more convection heating is added, especially on the soft-coated Low-E glass.

Production Data
Furnace Heating Rate
Clear vs. Soft coat low-e

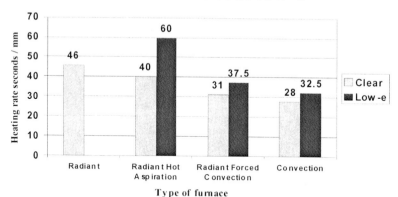

Figure 8: Furnace Heating Rates

SUMMARY

Looking to the future, the majority of flat glass manufacturers will continue to produce more and more products with high performance coatings. Many of these glasses will have coatings on both surfaces. Therefore, new glass tempering systems will be needed in order to process an ever-increasing number of glass products.

In the pre-processing area there will be challenges in the storage, selection and handling of glasses to be processed. After cutting glasses to size, the seaming, grinding, polishing and extra processing will become more difficult to accomplish without damaging the glass coatings. Conveying and washing of these products also will offer potential handling complications. One must keep in mind that a tempered glass product will be only as good as the glass going into the furnace.

To process the new generation of coated glasses, a full convection furnace will be needed. With this type of furnace, the furnace temperature set point can be reduced, and cycle times will be minimized. With less cycle time in the furnace, the optical quality of the product will improve. The best furnaces will have full convection heating on both the top and bottom surfaces of the glass, using hot fans to re-circulate the hot air. Ideally, the system would be designed to have separate convection control on both the top and bottom surfaces of the glass, utilizing a high-volume, low-velocity air stream. With this furnace design, heating by convection should be a minimum of 75% with a furnace energy efficiency of approximately 92%.

The basic glass tempering furnace will continue to be a roller hearth design utilizing fused silica rolls. Since the rolls are the source of roller wave distortion, tolerances will become tighter. For application in a full convection furnace, the use of hollow rolls would minimize conductive heating and allow more heat control on the bottom surface of the glass.

THE USABLE GLASS STRENGTH COALITION INITIATIVE TO PROVIDE FUNDING FOR FUNDAMENTAL RESEARCH IN GLASS STRENGTH

Robert Weisenburger Lipetz
Glass Manufacturing Industry Council
Columbus, Ohio USA

ABSTRACT

Glass use is limited by its properties of fragility and weight. The intrinsic strength of glass far exceeds its usable strength. Even modest gains in glass strength translate to industrial efficiencies. We have seen advances in understanding and application of technologies that result in increases in the usable glass strength for some isolated applications. Still, these technologies are inadequate. Although they increase the usable strength of some glass articles, they have not addressed the fundamental, root cause of low glass strength. It is becoming increasingly apparent, that fundamental breakthroughs will not come from individual company research. We have witnessed advancements in the experimental techniques university researchers can use to study glass strength. Programmatic support to rigorously put the techniques and the capabilities to use isn't currently available. Glass companies cannot independently support a fundamental research agenda to understand and significantly improve the usable strength of glass. However by working together with pooled funding and shared risk, the opportunity to significantly improve the usable strength of glass is achievable. Thus, the concept of developing a coalition to unite glass users, glass manufacturers, academicians, and government representatives, was born as an effort to begin crafting a research roadmap, identifying public and private funding, and negotiating a method of sharing information and collaboration.

INTRODUCTION

We, the Glass Manufacturing Industry Council (GMIC), present an account of efforts to establish a coalition to provide industry funding for fundamental research in glass strength. The GMIC has been acting as the coordinator of this recent effort to organize the glass industry around supporting fundamental research in glass strength as of the writing of this paper in fall 2012. GMIC is a non-profit trade association for the glass industry, founded in part, to act as a single voice for all segments of the glass industry. GMIC's mission is to facilitate, organize and promote the interests, economic growth and sustainability of the glass industry through education and cooperation on the areas of technology, productivity, innovation and the environment.

THEORETICAL VERSUS USABLE GLASS STRENGTH

Although glass is transparent, it is so ubiquitous, that we are often blind to how much we depend on glass products. We find it in architecture, in our homes, offices and public buildings. It is the preferred package for many food and beverages. Glass is a vital part of our renewable energy strategy. And the advent of next-generation chemistries and manufacturing processes has made glass the interface material on our modern electronic tools.

But still, glass products are often faulted for two key deficiencies; glass is heavy and breakable. As a result, glass is not always the material of choice when designing new products and applications. The question is, how do we as an industry, overcome these deficiencies?

We know that glass is actually very strong. The intrinsic strength of glass is stronger than steel (1). Glass also has a lower density than steel and aluminum. So if we were able to reach these high strengths and combined it with the lightweight factor, the number of new opportunities for glass applications would explode. We are familiar with the data shown in Table 1, which compares the theoretical strength of glass and the usable strength of glass. In the end, the brittle nature of glass is highly susceptible to the generation of fatal surface flaws during manufacturing and handling. As a result, the usable strength of glass has only reached a fraction of its potential and typical glass applications realize only 0.5% of the intrinsic strength.

TABLE 1. Thermal vs. Usable Strength of Glass

Condition of Glass	Strength (lb/in^2)
Theoretical/Lab Demonstrated	2,000,000
Pressed Articles	3,000-8,000
Blown Ware • Inner Surface	4,000-9,000 15,000-40,000
Drawn Tubing or Rod	6,000-15,000
Glass Fibers • Freshly Drawn • Annealed • Telecommunication	30,000-40,000 10,000-40,000 >100,000
Window Glass • LCD (0.65 mm) • Chemically Treated Cover Glass	8,000-20,00 45,000 100,000-200,000

Fortunately, we have seen a growth in understanding and application of technologies that result in dramatic increases in the usable glass strength for isolated applications. Improved processing techniques have enabled the creation of super-thin, high-strength flat glass used in applications like LCD panels. And secondary treatments like chemical strengthening have pushed the strength envelope even further. The processes produce very useful commercial applications of glass.

Even modest gains in glass strength translate to significant industrial efficiencies. For example, less raw material is needed to make the same products. Less energy is required, creating substantial savings and reducing emissions. Furnace campaigns are extended. Forming production requires less time. Stronger glass means lighter products. Transportation costs are reduced. Storage is more efficient. Stronger and lighter products multiply the uses available for the products and make them more efficient.

One well known strengthening technology is chemical treatment of glass and chemically tempered smartphone cover glass is becoming more common. But, chemical treatment is limited to certain applications, and because of the length of time it takes and other expenses, it isn't commercially viable for many uses. Thermal tempering of glass is used to produce safety windows in buildings and automobiles and allows us to walk 4000 ft above the floor of the Grand Canyon. Another way to strengthen glass is through lamination, such as bulletproof glasses.

Still, these technologies are inadequate. Although they increase the usable strength of some glass articles, they have not addressed the fundamental, root cause of low glass strength. Wouldn't it be better to understand the nucleation of flaws in glass, before applying these advanced tempering techniques to them? So the concept behind the UGSC is to develop a qualitatively stronger base of glass, and *then*, use strengthening techniques. This would make an enormous leap up in total glass strength.

THE CASE FOR AN INDUSTRY COALITION

Progress has been slow and it is becoming increasingly apparent, that fundamental breakthroughs will not come from individual company research. Historical gains have been small. Industry has closed over half of North American manufacturing plants in the past 20 years.

Members of the scientific and industrial community believe we are at a tipping point where we can exploit advancing fundamental laboratory techniques to better understand the origins of glass strength and engineer new products and processes to advance usable glass strength.

We have witnessed advancements in the experimental techniques university researchers can use to study glass strength. Over the last decade, we have seen the development of cutting edge research tools including:

- The two-point bend test for glass fibers. (see figure 1)
- Improved computer modeling of flaws and fracture surfaces, utilizing super computers in national labs and universities. (see figure 2)
- Aria stress birefringence technique, capable of studying the effect of impacts on the near surface region. (see figure 3)
- Advances in the use of atomic force microscopy to study flaw sites and the relaxation of indentation on the surface of glass to study cracks. (see figure 4)
- An increased understanding of surface phenomena including corrosion and the interaction of glass with coatings.

We now have a tool chest of new, robust techniques. Unfortunately, when it comes to glass strength, the funding to rigorously put the techniques and the capabilities to use isn't available. Unlike a few decades ago, glass research is spread out across many universities, laboratories, and private businesses. Thus, the concept of developing a coalition to unite glass users, glass manufacturers, academics, and government representatives, was born as an effort to begin crafting a research roadmap, identifying public and private funding, and negotiating a method of sharing information.

FIGURE 1. Two-point Bend Testing of Glass Fiber

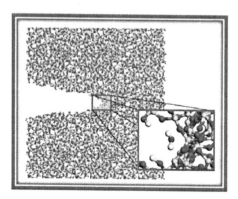

FIGURE 2. Molecular Modeling of Glass Structure

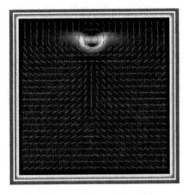

FIGURE 3. Abrio Stress Birefringence

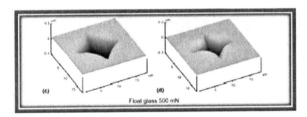

FIGURE 4. Atomic Force Microscope Images of Indentations?

We realized no one entity was going to be able to do it alone. Glass companies cannot independently support a fundamental research agenda to understand and significantly improve the usable strength of glass. However by working together with pooled funding and shared risk, the opportunity to significantly improve the usable strength of glass is achievable. We knew building the Usable Glass Strength Coalition (UGSC) wouldn't be easy and, so far, the journey has been three years in the making. One drawback was that few of the key participants had coalition-building experience. Furthermore, the participants were not, a homogeneous group. While the theoretical payoff of a successful coalition might be huge, there are also many risks. However, there is a good argument for private-sector participation in something like the UGSC. Most companies no longer perform fundamental research. The private sector's research is almost entirely focused on solving internal problems and manufacturing issues.

COALITION ORGANIZATION

A key event occurred in 2009, at the Pacific Rim Conference on Ceramic and Glass Technology in Vancouver, Canada, when a glass strength session brought together over 100 representatives from research and industry. The Vancouver session led to further meetings at Alfred University, Pennsylvania State University, and the American Society for Testing and Materials (ASTM) conference in Washington, D.C. We persisted in coming together as a team so that members of industry, government agencies and of universities could continue a conversation about this topic and how collectively we could push the research forward.

The initial phase of UGSC culminated in the 2010 Glass & Optical Materials meeting in Corning, N.Y., and obtained a small but important amount of seed funding. After that, we had a meeting at Coca-Cola, and then we were able to form a research roadmap at the American Ceramic Society's Glass & Optical Materials Division annual meeting in Savannah in 2011.

We crafted a mission statement. That mission statement led to a statement of our overarching objective, "To develop a precompetitive research program to identify critical parameters for improving the usable strength of glass." The key term here is "precompetitive." Competition among the companies is a very real concern, and by limiting the UGSC's work to "precompetitive" fundamental research, we think we can enable companies to work together as a consortium for the broader good. We broke down the UGSC objective into three parts. The first is to gain a fundamental understanding of methods for improving usable glass strength. The second part, develop and standardize new tools and testing methods, is critical. We all define strength differently and we don't use the same tools. The third part of the UGSC objective is to develop the next generation of glass technical experts and researchers.

An initial, informal organization of volunteers began with two groups: The Core Research Team (CRT) would take on the responsibility of defining the research roadmap that was necessary to achieve the long-term goal of significantly increasing the usable strength of glass. The CRT was comprised of leading researchers from universities, industry, and government agencies.

Its counterpart was the Strength Steering Team (SST). Their role was to define the coalition objectives and develop a model for funding the research roadmap developed by the CRT. The team is comprised of members of the "specialty," fiber and container-making interests. It has talked with several flat glass companies and believes there they will be more interested when the final membership documents and roadmap are ready.

The SST put together a seed-funding proposal. The idea was to accomplish the administrative and diplomatic tasks of addressing the structure of the coalition, drafting an operating agreement and a participation agreement, and, finally, creating a more robust research roadmap that we could use. The financial support would be used to fund the development of a formal operating agreement and a research roadmap. This commitment gave us the boost we needed and in September 2010, the UGSC members came together in Atlanta, GA to develop a framework for formalizing the coalition.

Coalition members agreed to an interim leadership that would oversee developing the legal structure of the organization. And then, at that point, the governance documents created, would dictate the election or appointment of new leaders. Doug Trenkamp of O-I took the lead position with substantial leadership sharing from others, including Elam Leed of Johns Manville, now replaced by Jeffery Shock, Louis Mattos at Coca-Cola, Alastair Cormack and myself at GMIC, as well as input from others like Richard Blanchard of Diageo.

Another problem was that the coalition seemed too abstract, lacking administration, management, intellectual property (IP) oversight mechanisms, etc. The coalition needed a firmer entity to be tied to, and the most logical choice was the GMIC. The GMIC already is a recognized glass industry representative and many of the companies participating in the UGSC discussions belong to the Council. GMIC agreed to allow UGSC to operate as a separate function within the council's existing structure. Under this arrangement, GMIC members do not automatically become UGSC members. Likewise UGSC members will not also have to be members of GMIC. GMIC still will focus on the industry and its initiatives, but the UGSC will be focused on funding universal research.

The UGSC has been incorporated as a for-profit limited liability corporation, wholly owned by the GMIC. The nature of the UGSC activities is spelled out in an operating agreement. The UGSC will have it's own independent board of directors that oversee research focus and management. The rights and dues of participating companies are detailed in participation agreements. A surprisingly enormous amount of work has been taking place over the past year to create an operating agreement that meets the objectives of the mission while addressing the concerns of coalition member companies.

Creating an agreement among diverse bodies of companies, some of whom are competitors, has proven to be more of a challenge than anticipated. Many issues are being worked out. Who can be members? What are levels and rights of members? Who gets a say in research projects and how much of a say? What happens when a new company enters the coalition or one leaves? How are dues structured? Are there legal issues around international membership? What are anti-trust concerns? How much funding is needed and how will it be spent? Key issues being defined in the operating agreement are:

1) It will be a stipulation of the research agreements that the research will be made public. Coalition members will have first access to the knowledge.
2) The very nature of pre-competitive research implies that at some point, competitive research will take place. All member companies will be able to independently take the results of the pre-competitive research to independently fund research leading to applied process IP.
3) Intellectual Property resulting from research will not be owned by the UGSC. This model of not retaining IP rights avoids the complications that could have occurred from a model that creates what amounts to a joint business venture.

Traditionally, when research is funded through universities, the university is assigned the IP and a royalty free license is granted to the sponsor. The leadership had initially worked on an operating agreement, in which generated knowledge would be owned by the members. Various mechanisms of licensing and control were created in greater and greater complexity. With the goal of creating a consensus among the coalition members, many iterations of drafts of the operating agreement were reviewed by various legal and IP experts, representatives of coalition companies' management and legal departments.

After reflecting on the long process, UGSC leaders concluded that it would be too difficult to gain consensus on an operating agreement that incorporated the ownership of IP by the participants. Rather than try to resolve conflicting IP interests, the interim leadership in December 2011 elected to take a new tack. They redrafted an operating agreement based on a model in which the generated IP would become public domain. UGSC has shared this new draft agreement with representatives of the potential member companies. At the time of this lecture, the drafts are still being reviewed by management, but the coalition leaders are optimistic that consensus on an operating agreement is nearing.

TABLE 2. Annual Participation Dues Structure

Participation Level	Sales/Purchase Metric	Annual Participation Fee	Voting Power
Gold	$\geq$ $1 billion	$40,000	4 votes
Silver	< $1 billion $\geq$ $100 million	$20,000	2 votes
Bronze	< $100 million	$10,000	1 vote

The Coalition participants are independent of the GMIC. The most current participation proposal is shown here. (see Table 2) In this proposal, the annual participation fee and board voting power will be determined on a sliding scale based on metrics associated with annual sales or purchases of the participating company.

Glass Manufacturers: Annual participation fee will be calculated based on annual sales.

Suppliers: Annual participation fee will be calculated based on annual sales to glass industry.

Users: Annual participation fee will be calculated based on annual purchase of glass products.

Universities: participation fee will be assigned at the Bronze Level.

Government Agency: Annual participation fee will be assigned at the Bronze Level.

Industry Consortium: Annual participation fee calculated based on the following formulas.

Gold Member = Number of participants × Silver Level

Silver Member = Number of participants × Bronze Level

The financial model is for UGSC to be a self-funded coalition for the first three years. The target would be to start off with six to ten industry participants, with a balance between manufactures and users. A minimum of $180,000 per year total commitment would be needed to commence operations. Then, after the initial three-year period, the goal is to have the UGSC's research advance to the point where the university representatives are in a better position to seek matching funds from government agencies because they will have built-up a research base and results to justify moving forward.

RESEARCH ROADMAP

Before companies formally can be approached to invest in the UGSC, the coalition needs to be able to say what the outcome will be three or five years down the road and show them the path we intend to take to get there. Along these lines, the coalition's CRT organized a research roadmap meeting in Savannah, Ga., in 2011, led by Alistair Cormack (Alfred University). The group affirmed the need to go all the way back to flaw generation in glasses. Additionally, we want to understand the relationship between crack initiation and contact damage. University researchers say they have studied defects and generally know where they come from. The missing piece is to understand how cracks and flaws link to manufacturing steps. We would want to understand the differences between mechanical and chemical damage. The UGSC also wants to look at the role of surface structure defects and reactive sites. There is a lot of work now on using coatings to protect glass or add functionality, but we need to understand what really the role of the coating is. To simplify the roadmap, the UGSC has developed a simple, two-step graphic. (see figure 5) The first step starts with understanding surface structure and chemistry. That is where the flaws of interest start. A component of this is understanding the effect of chemistry on surface structure and how surface structure is impacted by chemical and physical damage. And, that then, leads to reduction in strength. Therefore, the first goal of our research roadmap is to have the researchers tell us why glass shows a reduction in strength.

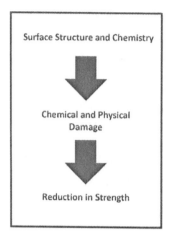
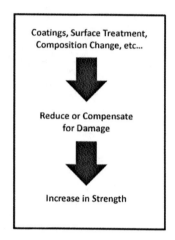

FIGURE 5. Schematic of the two-step approach

The roadmap's second step moves in a direction that is a little more application-specific. Admittedly, this may involve work that is no longer precompetitive research, but something that evolves into competitive research along with universities, and may be part of the coalition's function as the years go on.

Besides finalizing the operating agreement, the UGSC interim leaders have drafted a request for research proposals. The idea is to create a highly structured RFP that also will provide potential members with more specificity about the direction research will take. Currently, the leadership is collecting feedback from company representatives on the RFP.

The draft RFP is summarized below. Here are the basic questions to answer.

The Usable Glass Strength Coalition (UGSC) is seeking Research Proposals focused on gaining a deeper understanding of the relationship between Surface Structure, Surface Chemistry and the Strength of Glass. Specifically, the UGSC is seeking proposals to answer the fundamental question: Where and how do flaws nucleate in glass? This request will be limited to proposals that address the following research topics:

- What are the nature of the surface defect structures and reactive sites?
- What is the link between surface structure and adsorption?
- What are the changes in surface structure and surface chemistry which occur as glasses are aged from pristine conditions?
- What are the mechanisms behind the reduction in strength due to zero stress corrosion/aging?
- What are the differences between a fracture surface and a melt surface?

The first phase of research will concentrate on glass surface chemistry and structure. The second phase, then will look at mechanisms of chemical and physical damage.

Research Strategy: A number of industry and university experts have contributed to the creation of a research strategy. The overall focus of the research strategy can be Glass surface structure and chemistry, particularly as it relates to the susceptibility to chemical, thermal and physical damage, described in terms of improving the understanding of:

Phase 1: Glass surface structure and chemistry, particularly as it relates to the susceptibility to chemical, thermal and physical damage.

Phase 2: Mechanisms of chemical and physical damage (flaw generation) as they relate to usable strength reduction.

At this time, proposals should focus on Phase 1 topics

Possible Approaches:

- The research effort should focus on boron-free e-glasses and soda-lime silica glasses.
- Use of fibers as a model system
- Computer Modeling of glass surfaces and their relationship to glass chemistry.
- Surface structure characterization
- Surface chemistry characterization
- Adsorption mechanisms as they relate to surface structure and chemistry

To be clear, the areas that the coalition is not interested in pursuing in the first phase are stated. So, areas of research like coatings, strengthening techniques and crack propagation, would not be of interest initially. Approaches not of initial first round focus:

- Damage mechanisms due to chemical interactions
- Damage mechanisms due to physical interactions
- Strengthening techniques
- Coating technologies
- Propagation of cracks
- Mechanical properties measurements
- Metallic glasses

LESSONS LEARNED

We have come a long way in this attempt to provide industry funded fundamental research in glass strength. No one anticipated that it would be this challenging or take this long. It still remains to be seen how successful this effort will be. We have learned much from this attempt. Trying to form a research coalition of industry partners is hard work. In these economically challenging times, it is difficult to justify investment in fundamental research that may not provide an immediate benefit. It is not clear how a coalition that has carte blanche to research in whatever areas it is interested in, will conflict with the proprietary technologies of the participating companies.

Although most of the participating companies belong to multiple research consortia, they find it much easier to join coalitions, than to form new coalitions. They fear it will become a joint venture, which can become legally complicated. By putting UGSC under GMIC to form this entity, companies have an established organization they can join.

The interest in improving the strength of glass is a hot topic for industry, and some governments are starting to catch on to what sweeping changes could be made if there are breakthroughs. It is a confirmation of the UGSC's vision that the German Science Foundation (DFG) agreed to 11 million Euros of funding for research on ultrastrong glasses. And it is a testament to the UGSC mission that glass industry companies have provided seed funding and donated hundreds of hours of time towards achieving this goal. The supporters of the UGSC know the critical need for glass strength breakthroughs, and are determined to work to find the path forward.

ACKNOWLEDGEMENTS:

Based on a presentation by Louis Mattos, Jr., The Coca Cola Company, at the American Ceramic Leadership conference in August of 2011.

Table 1 - John T. Brown and Suresh Gulati, Corning

Table 2 – Robert Weisenburger Lipetz, Glass Manufacturing Industry Council

Figure 1- Richard K. Brow, Missouri University of Science & Technology

Figure 2 - Elam Leed, Johns Manville

Figures 3 & 4 - Charles Kurkjian, University of Southern Maine

Figure 5 – Alastair Cormack, Alfred University

REFERENCES:

1. Smith, W.A. and Michalske, T.M. DOE contract #DE-AC04-0DPOO789, (1990).

2. J. von Kowalski, "Fetigkieit des Glases," Ann. d. Phys. u. Chem. 36 307 (1889) referenced in Murgatroyd, J. Soc. Glass Tech. 28 406-31T (1944).

3. C. Brodmann, "EinigeBeobachtungenUber die Festigkeit von Glasstaben," Nachr. Konigl. Ges. Wiss., 44 44–58 (1894) referenced in Murgatroyd, J. Soc. Glass Tech. 28 406-31T (1944).

4. A. A. Griffith, "The Phenomena of Rupture and Flow in Solids," Philos. Trans. R. Soc. Lond., 221 163–198 (1920).

5. J. E. Gordon, The New Science of Strong Materials, Penguin Books, New York, 1976.

6. L. Grenet, Bull. Soc. Encour. Ind. Nat. [Ser. 5] 4 838 (1899). For translations of the paper see: F.W. Preston, "The Time Factor in the Testing of Glassware," J. Am. Ceram. Soc. 18 220–224 (1935), and Glass Ind., 15 [11] 277–80 (1934).

7. S. W. Freiman, S. M. Wiederhorn, and J. J. Mecholsky Jr, "Environmentally Enhanced Fracture of Glass: A Historical Perspective," J. Am. Ceram. Soc., 92 1371–1382 (2009).

8. F. W. Preston, "The Structure of Abraded Glass Surfaces," Trans. Optical Soc., 23 141–164 (1921).

9. F. W. Preston, "A Study of the Rupture of Glass," J. Soc. Glass. Tech., 10 234 (1926).

10. L. H. Milligan, "The Strength of Glass Containing Cracks," J. Soc. Glass Technol., 13 351–360 (1929).

11. F. W. Preston, "The Time Factor in Testing Glass-Ware," J. Am. Ceram. Soc., 18 220 (1935).

12. L. V. Black, "Effect of the Rate of Loading on the Breaking Strength of Glass," Bull. Am. Ceram. Soc., 15 274–275 (1936).

13. T. C. Baker and F. W. Preston, "Fatigue of Glass Under Static Loads," J. Appl. Phys. 17, 170–178 (1946).

14. J. L. Glathart and F. W. Preston, "The Fatigue Modulus of Glass," J. Appl. Phys., 17 189–195 (1946).

15. C. Gurney and S. Pearson, "The Effect of the Surrounding Atmosphere on the Delayed Fracture of Glass," Proc. Phys. Soc. B, 62 469–476 (1949).

16. E. Orowan, "The Fatigue of Glass Under Stress," Nature, 154 341–343 (1944).

17. C. Gurney, "Delayed Fracture in Glass," Proc. Phys. Soc. (London), 59 169 –185 (1947).

18. D. A. Stuart and O. L. Anderson, "Dependence of Ultimate Strength of Glass Under Constant Load on Temperature Ambient Atmosphere, and Time," J. Am. Ceram. Soc., 36 416–424 (1953).

19. M. Watanabe, R. V. Caporali, and R. E. Mould, "The Effect of Chemical Composition on the Strength and Static Fatigue of Soda-Lime Glass," Phys. Chem. Glasses, 2 12–23 (1961).

20. G. R. Irwin "Fracture;" Handbuch der Physik, Vol. 6. ed., S. Flugge. Springer-Verlag, Berlin-Heidelberg, 551–590, 1958.

21. D. M. Kulawansa, L. C. Jensen, S. C. Langford, and J. T. Dickinson, "STM Observations Of The Mirror Region Of Silicate Glass Fracture Surfaces," J. Mater. Res., 9 476–485 (1994).

22. R. L. Smith III, J. J. Mecholsky Jr, and S. W. Freiman, "Estimation of Fracture Energy From the Work of Fracture and Fracture Surface Area: I. Stable Crack Growth," Int. J. Fract., 156 97–102 (2009).

23. J. J. Mecholsky Jr, S. W. Freiman, and R. W. Rice, "Effect of Grinding on Flaw Geometry and Fracture of Glass," J. Am. Ceram. Soc., 60 [3–4] 114–117 (1977).

24. J. J. Mecholsky Jr, S. W. Freiman, and R. W. Rice, "Fracture Surface Analysis of Ceramics," J. Mat. Sci., 11 1310–1319 (1976).

25. N. Shinkai, R. C. Bradt, and G. E. Rindone, "Elastic Modulus and Fracture Toughness of Ternary PbO-ZnO-B2O3 Glasses," J. Am. Ceram. Soc., 65 123–126 (1982).

26. R. J. Eagan and J. C. Swearengen, "Effect of Composition on the Mechanical Properties of Aluminosilicate and Borosilicate Glasses," J. Am. Ceram. Soc., 61 27–30 (1978).

27. E. Vernaz, F. Larche, and J. Zarzycki, "Fracture Toughness-Composition Relationship in Some Binary and Ternary Glass Systems," J. Non-Cryst. Solids, 37 359–365 (1980).

28. S. W. Freiman, T. L. Baker, and J. B. Wachtman Jr, "A Computerized Fracture Mechanics Database for Oxide Glasses," Bull. Am. Ceram. Soc., 64 1452–1455 (1985).

29. C. R. Kennedy, R. C. Bradt, and G. E. Rindone, "Fracture Mechanics of Binary Sodium Silicate Glasses"; Fracture Mechanics of Ceramics, Vol. 2, eds., R. C. Bradt, A. G. Evans, D. P. H. Hasselman and F. F. Lange. Plenum Press, New York, 883–893, 1973.

30. J. A. Salem and M. G. Jenkins, "Estimating Bounds on Fracture Stresses Determined From Mirror Size Measurements," J. Am. Ceram. Soc., 85 706 –708 (2002).

31. H. P. Kirchner, "The Strain Intensity Criterion for Crack Branching in Ceramics," Eng. Fract. Mech., 10 283–288 (1978).

32. J. J. Mecholsky Jr, A. C. Gonzalez, and S. W. Freiman, Fractographic Analysis of Delayed Failure in Soda Lime Glass," J. Am. Ceram. Soc., 62 577–580 (1979).

33. J. C. Conway Jr, and J. J. Mecholsky Jr, "Use of Crack Branching Data for Measuring Near-Surface Residual Stresses in Tempered Glass," J. Am. Ceram. Soc., 72 1584–1587 (1989).

34. D. B. Marshall, B. R. Lawn, and J. J. Mecholsky Jr, "Effect of Residual Contact Stresses on Mirror/FlawSize Relations," J. Am. Ceram. Soc., 63 7– 8 (1980).

35. J. J. Mecholsky, D. E. Passoja, and K. S. Feinberg-Ringel, "Quantitative Analysis of Brittle Fracture Surfaces Using Fractal Geometry," J. Am. Ceram. Soc., 72 60–65 (1989).

36. E. Bouchaud, "Scaling Properties of Cracks," J. Phys. Condens. Matter, 9 4319–4345 (1997).

37. J. K. West, J. J. Mecholsky Jr, and L. L. Hench, "The Quantum and Fractal Geometry of Brittle Fracture," J. Non-Cryst. Solids, 260 99–108 (1999).

38. J. J. Mecholsky Jr, and S. W. Freiman, "Relationship Between Fractal Geometry and Fractography," J. Am. Ceram. Soc., 74 [12] 3136–3138 (1991).

39. D. E. Passoja, "Fundamental Relationships Between Energy and Geometry in Fracture. Fractography: A Quantitative Measure of the Fracture Process;" Ceramic Transactions 17 Fractography of Glasses and Ceramics II, eds. V. D. Frechette and J. Varner. American Ceramic Society, Columbus, 101 –126, 1991.

40. S. M. Wiederhorn, "Influence of Water Vapor on Crack Propagation in Soda–Lime Glass," J. Am. Ceram. Soc., 50 [8] 407–414 (1967).

41. K. Jakus, J. E. Ritter Jr, and J. M. Sullivan, "Dependency of Fatigue Predictions on the Form of the Crack Velocity Equation," J. Am. Ceram. Soc., 50 [8] 407–414 (1967).

42. S. W. Freiman, "Effect of Alcohols on Crack Propagation in Glass," J. Am. Ceram. Soc., 64 372–374 (1981).

43. S. M. Wiederhorn and H. Johnson, "Effect of pH on Crack Growth in Soda Lime Silica Glass," J. Am. Ceram. Soc., 56 192–197 (1973).

44. G. S. White, S. W. Freiman, S. M. Wiederhorn, and T. D. Coyle, "Effects of Counterions on Crack Growth in Vitreous Silica," J. Am. Ceram. Soc., 70 [12] 891–895 (1987).

45. S. M. Wiederhorn and L. H. Bolz, "Stress Corrosion and Static Fatigue of Glass," J. Am. Ceram. Soc., 53 [10] 543–548 (1970).

46. S. M. Wiederhorn, H. Johnson, A. M. Diness, and A. H. Heuer, "Fracture of Glass in Vacuum," J. Am. Ceram. Soc., 57 336–341 (1974).

47. T. I. Suratwala and R. A. Steele, "Anomalous Temperature Dependence of sub-Critical Crack Growth in Silica Glass," J. Non-Cryst. Solids, 316 174–182 (2003).

48. G. A. Fisk and T. A. Michalske, "Laserbased and Thermal Studies

of Stress Corrosion in Vitreous Silica," J. Appl. Phys., 58 2736–2743 (1985).

49. B. R. Lawn, "Physics of Fracture," J. Am. Ceram. Soc., 66 [2] 83–91 (1983).

50. S. M. Wiederhorn, S. W. Freiman, E. R. Fuller Jr, and C. J. Simmons, "Effect of Water and Other Dielectrics on Crack Growth," J. Mater. Sci., 17 3460–3478 (1982).

51. S. W. Freiman, "Temperature Dependence of Crack Propagation in Glass in Alcohols," J. Am. Ceram. Soc., 58 340–341 (1975).

52. R. J. Charles and W. B. Hillig, Symposium on Mechanical Strength of Glass and Ways of Improving It, Florence, Italy, September 25–29 1961,

511–527, Union Scientifique Continentale du Verre, Charleroi, Belgium, 1962.

53. T. A. Michalske, "The Stress Corrosion Limit: Its Measurement and Implications;" Fracture Mechanics of Ceramics, Vol. 5, eds., R. C. Bradt, A. G. Evans, D. P. H. Hasselman and F. F. Lange. Plenum Press, New York, 277–289, 1977.

54. E. Gehrke, C. H. Ullner, and M. Ha˝hnert, "Fatigue Limit and Crack Arrest in Alkali-Containing Silicate Glasses," J. Mater. Sci., 26 5445–5455 (1991).

55. C. J. Simmons, and S. W. Freiman, "Effect of Corrosion Processes on Subcritical Crack-Growth in Glass," J. Am. Ceram. Soc., 64 [11] 683–686 (1981).

56. T. A. Michalske and S. W. Freiman, "A Molecular Mechanism for Stress Corrosion in Vitreous Silica," J. Am. Ceram. Soc., 66 [4] 284–288 (1983).

57. T. A. Michalske and B. C. Bunker, "Slow Fracture Model Based on Strained Silicate Structures," J. Appl. Phys., 56 2686–2693 (1984).

58. B. R. Lawn and T. R. Wilshaw, Fracture of Brittle Solids, 168–172, Cambridge University Press, Cambridge, England, 1972. www.ceramics.org/IJAGS The Fracture of Glass 103

59. T. A. Michalske and B. C. Bunker, "Steric Effects in Stress Corrosion Fracture of Glass," J. Am. Ceram. Soc., 70 780–784 (1987).

60. S. W. Freiman, G. S. White, and E. R. Fuller Jr, "Environmentally Enhanced Crack Growth in Soda-Lime Glass," J. Am. Ceram. Soc., 69 38–44 (1985).

61. G. S. White, D. C. Greenspan, and S. W. Freiman, "Corrosion and Crack Growth in 33% Na2O-67% SiO2 and 33% Li2O-67% SiO2 Glasses," J. Am. Ceram. Soc., 69 [1] 38–44 (1986).

62. C. R. Kurkjian, P. K. Gupta, and R. K. Brow, "The Strength of Silicate Glasses: What Do We Know, What Do We Need to Know?," Int. J. App. Glass Sci., 1 27–37 (2010).

63. M. Muraoka and H. Abe´, "Subcritical Crack Growth in Silica Optical Fibers in a Wide Range of Crack Velocities," J. Am. Ceram. Soc., 79 51–57 (1996).

64. M. J. Matthewson and C. R. Kurkjian, "Environmental Effects on the Static Fatigue of Silica Optical Fiber," J. Am. Ceram. Soc., 71 177–183 (1988).

65. J. J. Mecholsky Jr, S. W. Freiman, and S. M. Morey, "Fractographic Analysis of Optical Fibers," Bull. Am. Ceram. Soc., 56 1016–1017 (1977).

66. T. A. Michalske, W. L. Smith, and B. C. Bunker, "Fatigue Mechanisms in High-Strength Silica-Glass Fibers," J. Am. Ceram. Soc., 74 1993–1996 (1991).

67. W. Wong-Ng, G. S. White, and S. W. Freiman, "Application of Molecular Orbital Calculations to Fracture Mechanics: Effect of Applied Strain on Charge Distribution in Silica," J. Am. Ceram. Soc., 75 [11] 3097–3102 (1992).

68. C. G. Lindsay, G. S. White, S. W. Freiman, and W. Wong-Ng "Molecular-Orbital Study of an Environmentally Enhanced Crack-Growth Process in Silica," J. Am. Ceram. Soc. 77 [8] 2179–2187 (1994).

69. J. E. Del Bene, K. Runge, and R. J. Bartlett, "A Quantum Chemical Mechanism for the Water-Initiated Decomposition of Silica," Comput. Mater. Sci., 27 102–108 (2003).

70. J. K. West and L. L. Hench, "The Effect of Environment on Silica Fracture: Vacuum, Carbon Monoxide, Water and Nitrogen," Phil. Mag. A, 77 [1] 85–113 (1998).

71. T. Zhu, J. Li, X. Lin, and S. Yip, "Stress-Dependent Molecular Pathways of Silica-Water Reaction," J. Mech. Phys. Solids, 53 1597–1623 (2005).

72. S. N. Crichton, M. Tomozawa, J. S. Hayden, T. I. Suratwala, and J. H. Campbell, "Subcritical Crack Growth in a Phosphate Laser Glass," J. Am. Ceram. Soc., 82 [11] 3097–3104 (1999).

73. S. W. Freiman and T. L. Baker, "Effects of Composition and Environment on the Fracture of Fluoride Glasses," J. Am. Ceram. Soc. 71 [4] C-214–C-216 (1988).

74. J. J. Mecholsky Jr, R. W. Rice, and S. W. Freiman, "Prediction of Fracture Energy and Flaw Size in Glasses From Mirror Size Measurements," J. Amer. Ceramic Soc., 57 440–443 (1974).

75. S. W. Freiman and J. J. Mecholsky Jr, "The Fracture Energy of Brittle Crystals," J. Mat. Sci., 45 4063–4066 (2010).

76. F. Ce'larie' et al. "Glass Breaks Like Metals, but at the Nanometer Scale," Phys. Rev. Let., 90 [7] 075504 (2003).

77. S. Prades, D. Bonamy, D. Dalmas, E. Bouchaud, and C. Guillot, "Nano-Ductile Crack Propagation in Glasses Under Stress Corrosion: Spatiotemporal Evolution of Damage in the Vicinity of the Crack tip," Int. J. Solids Struct., 42 [2] 637–645 (2005).

78. J.-P. Guin and S. M. Wiederhorn, "Fracture of Silicate Glasses: Ductile or Brittle?," Phys. Rev. Let., 92 [21] 215502 (2004).

79. J.-P. Guin and S. M. Wiederhorn, "Surfaces Formed by Subcritical Crack Growth in Silicate Glasses," Int. J. Frac., 140 15–26 (2006).

80. K. Han, M. Ciccotti, and S. Roux, "Measuring Nanoscale Stress Intensity Factors With an Atomic Force Microscope," EPL, 89 66003 (2010).

81. M. Tomozawa, W.-T. Han, and W. A. Lanford, "Water Entry Into Silica Glass During Slow Crack Growth," J. Am. Ceram. Soc., 74 [10] 2573–2576 (1991).

82. S. M. Wiederhorn, T. Fett, G. Rizzi, S. Funfschilling, M. J. Hoffmann, and J.-P. Guin, "Effect of Water Penetration on the Strength and Toughness of Silica Glass," J. Am. Ceram. Soc. (submitted).

Refractories

PERFORMANCE OF FUSION-CAST AND BONDED REFRACTORIES IN GLASS MELTING FURNACES

Amul Gupta and Kevin Selkregg
RHI Monofrax

Matthew Wheeler and Goetz Heilemann
RHI US Ltd.

ABSTRACT

Fusion-cast refractories continue to be the preferred choice for lining of glass melting furnaces. However, the growing imperative to reduce manufacturing cost has caused the glass industry to take more interest in the bonded (also known as sintered) refractories. Such refractories can be made by different methods, such as traditional dry pressing, iso-pressing and vibro-casting. This paper compares the performance of fusion-cast and bonded refractories in laboratory tests.

INTRODUCTION

The pressure to reduce manufacturing cost of glass and increase furnace life while still making high quality glass continues to rise. It is not easy to achieve these often conflicting goals through selection of lower cost refractory materials alone. That said, some segments of glass industry, such as container glass, have aggressively pursued lower-cost refractory materials. Examples of such materials include fusion-cast refractories made in countries with lower labor cost, for example China and India; and bonded refractories. As the glass industry strives to meet this challenge, it is important for the refractory suppliers to provide trustworthy data on performance of their refractories.

This paper will attempt to provide some insights into the testing and evaluation process for fusion-cast and bonded materials. Since the primary goal of a refractory material, especially in glass contact application, is to last for long time, corrosion testing becomes a key test for both refractory suppliers and the glass companies. The pros and cons of different corrosion test methods will be discussed.

Two refractory groups will be considered in this paper, AZS based and alumina based. For AZS based products, refractories made with fusion-casting, dry pressing, and vibro-casting processes will be considered. The Alumina based products include only fusion-cast and vibro-cast refractories. Although the bulk of the paper will focus on performance of these materials in a corrosion test, some information on glass defect potential will also be provided.

AZS BASED REFRACTORIES

Table I shows physical and chemical properties of AZS products considered in this study for superstructure and glass contact applications (for example, breast walls, port arches, jambs, melter pavers, and fore hearth channels). It is clear that the bonded and vibro-cast refractories have lower bulk density, higher apparent porosity, lower thermal conductivity and lower thermal expansion, when compared with the fused-cast refractories. These differences exist despite similar chemistry, for example zirconia content in the range 25-34 weight %.

Figure 2 contains images that reveal the microstructural differences between the bonded, vibro-cast, and fused-cast AZS refractories. The microstructure of the fusion-cast AZS material is a mixture of corundum and zirconia co-precipitates, isolated zirconia dendrites, and a sodium alumino silicate glass phase comprising one third of the AZS body by volume. Fusion-cast AZS refractories are not homogeneous in their chemistry and microstructure within one block. This results from faster cooling rate in the outer region of the block, and slower cooling in the interior region. It is typical to see a big range in grain size (less than 50microns to more than 1000 microns) within one block.

Table I : Key properties of fused-cast and bonded AZS refractories

AZS Products for Superstructure & Glass Contact Application								
Breastwalls, Port Arches. Jambs, Melter Pavers, Forehearth Channels								
Process	ZrO2	Major Phases	Bulk Density	Apparent Porosity	Thermal Conductivity W/m K			Thermal Expansion
	wt%		g/cc	%	1500C	1200C	1000C	% at 1400C
Fused-cast	32 - 34	ZrO2 alpha Alumina Glass phase	3.44 - 3.68	< 2	2	2.56	2.5	0.7
Vibro-cast	33	ZrO2 alpha Alumina Mullite	3.1	18	1.95	2.14	1.82	0.56
Pressed & Sintered	25 - 34	ZrO2 alpha Alumina Mullite	3.12 - 3.21	15 - 17			1.9	0.64 - 0.68

In general, the typical bonded and vibro-cast AZS materials consist of coarse-grained grog bonded by a fine-grained matrix of similar chemical and phase constituents. Typical grains are 500microns and higher. The large grog grains can be either previously fused or grown in-situ during firing. For example, grog containing co-precipitated zirconia/mullite grains can either be made from a fusion process, or grown in-situ during firing of brick that contained calcined alumina and Zircon sand as starting raw materials. The morphology of the two primary phases, Zirconia and mullite, can be quite different depending on the starting raw materials.

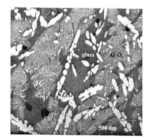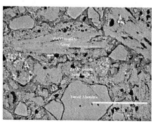

Figure 1: Microstructure of fused-cast AZS (left), sintered AZS (middle) and vibro-cast AZS (right)

The bonded and vibro-cast AZS refractories contain little or no glass phase. The glass phase could be considered the "bonding" structure holding the fused-cast AZS together while mullite forms the bond in the bonded and vibro-cast AZS refractories.

Corrosion from Contact with Glass Melt

Glass contact corrosion of refractory samples is commonly measured using ASTM C-621 static finger corrosion test or a similar test. In this test, 0.5" diameter by 4" long cylindrical finger shaped sample are immersed in a platinum crucible containing glass cullet of choice, and tested at desired temperature for a period of time. Typically, this test shows maximum corrosion at the glass melt line (also known metal line) in soda lime type glasses, and much lower corrosion below the metal line. When selecting refractories for bottom paving application, some refractory suppliers have used below metal line corrosion data on finger shaped samples.

Refractory corrosion can be measured simply as depth of cut at and below the metal line on corroded samples sawed in half along their length. Alternatively, corrosion can be measured as area lost at and below metal line using image analyzer software. The latter technique may be a better choice when refractory corrosion of a finger shaped sample is not uniform along its circumference. Figure 2 is a bar chart comparing the performance of a 34% ZrO_2 fused-cast AZS sample with a bonded AZS sample, following a corrosion test at 1427°C for 72h using soda lime container glass. It is clear that the bonded AZS sample showed twice the rate of corrosion of fusion-cast AZS sample at the metal line. However, below the metal line both samples actually showed a gain in volume despite draining the adhering glass. Furthermore, there appears to be more growth in the fusion-cast AZS sample than in the bonded AZS sample. The growth in the below metal line area of the finger samples made it difficult to accurately compare the performance of these materials.

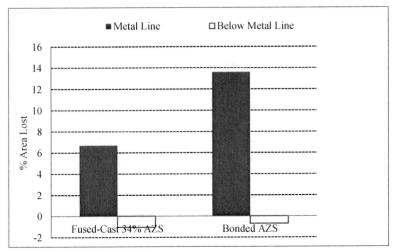

Figure 2: Comparison of fusion-cast vs. bonded AZS

In order to understand the mechanism of corrosion of bonded AZS below the metal line, the microstructure was characterized. Figure 3 shows the virgin bonded AZS microstructure on left and the corroded sample on right. The mullite phase (both as primary grog and the bond) was found to be disassociated throughout the finger sample. This was confirmed in an XRD scan shown in Figure 4. No diffraction evidence was found for primary mullite. This is due to easy diffusion of Na_2O, and to a lesser degree CaO, from container glass melt into the refractory sample through high level of open pores. In presence of Na_2O, mullite phase is not stable and breaks down to form alumina crystals and a sodium calcium silicate glass phase. In contrast, the fusion-cast AZS sample showed characteristic changes in its structure, i.e. diffusion of Na_2O in the glassy phase of AZS, resulting in dissolution of some crystalline Al_2O_3 in the glass phase. Thus, the fusion-cast AZS sample develops a viscous layer that acts as a passivating layer and slows down rate of corrosion below the metal line.

 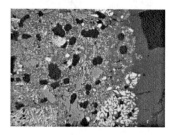

Figure 3: As-is bonded AZS refractory (left); corroded bonded AZS (right)

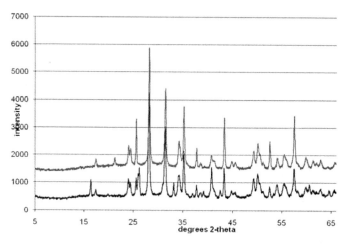

Figure 4: XRD scans of as-is untested bonded AZS (top spectra), and corroded bonded AZS (bottom spectra). Corroded sample does not show mullite peaks.

Because fused-cast AZS develops a boundary layer at the glass-refractory interface, whereas bonded AZS do not, in the next test finger samples of five different grades of bonded AZS, made by different suppliers, were tested for corrosion. These refractories varied in ZrO_2 concentration from about 26% to 34% and the apparent porosity varied from 9% to 15%.

Figure 5 shows corrosion at metal line and below metal line. Keeping in mind that bonded AZS are used in the bottom paving application, the metal line corrosion may not be very pertinent. That said, sample number 4 that had the lowest apparent porosity showed lower rate of corrosion at metal line. However, this sample showed the highest corrosion below the metal line. All 5 samples show a negative number for the % area lost due to corrosion below the metal line. In other words, the samples actually gained area, probably due to physical penetration of glass in the pores. Sample 4 showed largest gain in area despite having lowest apparent porosity of all the 5 samples tested. It can be concluded that the static finger corrosion test proved to be of limited use in comparing different bonded AZS samples.

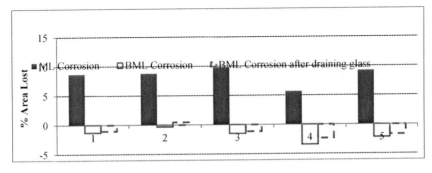

Figure 5: Bonded AZS sample numbers shown on horizontal axis

In order to better understand the below melt line corrosion behavior, a crucible test was performed next. The crucibles had approximately 3" inside diameter and approximately 2.25" depth of cavity. Therefore, this sample represents a large area and is more representative of coarse-grained refractory products. Figure 6 shows the bonded and fusion-cast AZS samples, cut in half, following a corrosion test at 1427°C for 96h. Although the bonded AZS sample showed more material loss at the metal line compared to fusion-cast AZS sample, the bottom surface of both crucibles showed no material loss and similar visual appearance.

Next, crucible corrosion test was performed on 4 different bonded AZS, made by different suppliers, using the same test conditions as described above. All the bonded materials tested showed similar visual appearance at the bottom of the crucibles (see Figure 7). In order to quantitatively differentiate bonded AZS made by different suppliers, the authors believe that crucible tests need to be performed for longer (than 4 days) test period. If there is very little physical loss of material due to corrosion, then it is necessary to utilize microscopy to determine and quantify changes in the microstructure of the refractory materials tested.

Figure 6: Bonded AZS crucible (left), fused-cast AZS (right) following corrosion test

Figure 7: Different bonded AZS crucibles following corrosion test. Clockwise from top left, sample 5, sample 4, sample 2, and sample 1. Same sample designation also applies in Figure 5.

Corrosion from Batch Carryover and Vapors

It is well known that corrosion of refractories used in the superstructure of glass melting furnaces can occur due to reaction with components of raw batch (also known as batch carryover) such as silica sand and soda ash; and also from vapor phase species, such as NaOH. While there is a standard test available from ASTM for corrosion of refractories from vapors (C987), there is no standard test available for testing corrosion from batch carryover. The ASTM standard, C987, requires the use of either alumina or platinum crucible for melting batch components that produce vapors, such as sodium carbonate. For the purpose of this study, the authors chose to prepare crucibles directly from the fusion-cast AZS and vibro-cast AZS products.

As is well known, fused-cast AZS has been primary material of choice thus far in the superstructure of most glass melting furnaces, especially in the batch charging area where carryover is present. In recent years, many container furnaces have been testing vibro-cast AZS products in port arches, breast walls, burner blocks, and in some cases even as feeder channels. In order to test the corrosion from glass melt, batch carryover, and vapors, we prepared crucibles from vibro-cast and fused-cast AZS products. In addition, lids were prepared from both types of products. Two separates tests were performed. In one test, soda-lime container glass cullet was placed in the crucibles and lids were placed on top of the crucibles. This test allowed us to use the crucibles for measuring glass melt corrosion, and the lids for corrosion from vapors. In the second test, crucibles were filled with soda-lime container glass raw batch. Raw batch was placed almost to the top of the crucible and lids were placed on top (Figure 8). The close proximity of the lids to the batch made sure that during heating up of the test furnace, there would be contact between the batch components and the lids, to simulate batch carryover conditions in glass furnaces. Both tests were performed at 1427°C for 96h.

Figure 8: Vibro-cast AZS crucible on left, Fused-cast AZS on right with corresponding lids

After the test, both the crucibles and the lids were examined visually for evidence of physical loss of material. In order to get a good look at the bottom of the crucibles, an oil of certain refractive index was poured on top of the glass. The results were quite revealing for the crucibles tested with batch in that the vibro-cast AZS sample showed significant *pitting* of the bottom, while the fused-cast crucible bottom was smooth (Figure 9). Following this observation, the crucibles were cut in half to allow an inspection of the metal line and below metal line corrosion. It was immediately clear that the height of the glass pool was different in the crucibles. The glass height was 12 – 20% lower in the vibro-cast AZS tested crucibles than the fused-cast AZS tested crucibles (Figure 10). Given that all crucibles had exactly the same dimensions and the quantity of glass cullet used was also the same, the lower height of the glass in the case of vibro-cast AZS implied *absorption* of glass within the porosity of the crucible. The fused-cast AZS material showed 30-35% less melt line corrosion then the vibro-cast AZS crucible (Figure 11).

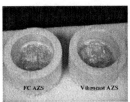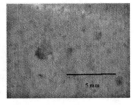

Figure 9: Batch carryover test. The photo in the middle shows the crucibles after the test. Photo on left shows the bottom of the FC AZS crucible. The photo on right shows the bottom of the Vibro-cast AZS crucible.

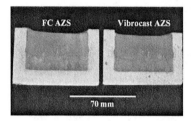

Figure 10: Left photo shows crucibles following test with raw batch; right photo shows crucibles following test with cullet. Note lower glass height in vibro-cast crucibles.

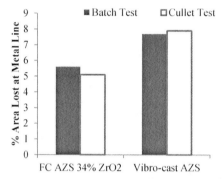

Figure 11: Corrosion of FC AZS and Vibro-cast AZS following crucible tests

The visual inspection of both types of AZS lids from test with soda-lime cullet showed little or no change in the quality of the exposed surface relative to non-tested surfaces. However, the vibro-cast AZS sample from the batch test showed significant pitting on its surface (Figure 12). In order to quantify the changes in the lids, a piece taken from the center of the lids was tested for chemistry and apparent porosity. Table II shows the apparent porosity of the lids from both tests. The fused-cast AZS lids showed similar porosity as seen in untested material. The vibro-cast AZS lid from the vapor test showed similar level of porosity as seen in untested material. However, the vibro-cast AZS lid from the batch test showed a significant reduction in porosity after the test. Figure 13 shows the chemistry profile of SiO_2 and Na_2O in the vibro-cast AZS lids after the test. The vibro-cast lid from the batch test showed ~ 10 – 20% more corrosion than the lid from the cullet test. The chemical change and the reduction in porosity in the vibro-cast AZS lid following the batch test are indicative of the absorption of batch within the high porosity of vibro-cast AZS.

  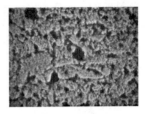

Figure 12: Left photo shows crucibles with lids before the test with raw batch. Middle photo shows FC AZS lid after the test, right photo shows Vibro-cast AZS lid after the test.

Table II: % Porosity of Exposed & Unexposed Lids

	Vibro-cast AZS	FC AZS
As-is	19.21	1.51
Cullet Test	19.48	1.53
Batch Test	13.91	1.39

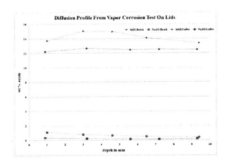

Figure 13: Chemical changes in the vibro-cast AZS lids following batch and cullet tests

Glass Defect Potential

A seed and blister potential test was carried out on fused-cast and vibro-cast AZS crucibles using soda lime glass at 1300°C for 72h. Figure 14 shows photos of a section of the crucibles. The bubbles that were generated during the test were attached to the refractory wall surfaces and not found in the bulk glass. Both types of AZS showed similar level of bubbles.

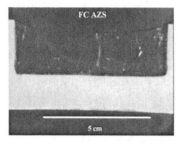

Figure 14: Sections of crucibles following seed and blister test. Bubbles seen adhering to crucible bottom – glass interface

SUMMARY
Commonly used finger test is not the best way to test bonded refractories used for glass furnace bottom paving application, even when below metal line data is used. Crucible tests provide a better means for comparing the performance of bonded and fused-cast AZS refractories. While fused-cast AZS refractories are expected to undergo little change due to their low level of open porosity and formation of passivation layer at the glass – refractory interface, both bonded and vibro-cast AZS are expected to show greater degree of chemical alteration due to higher level of porosity that allows penetration of glass melt. Short-term tests (96h) performed in this study did not show visual difference between 4 different bonded AZS products, made by different suppliers. Longer term test may be necessary to differentiate not only between fused-cast AZS and bonded AZS, but also different bonded AZS products. Vibro-cast AZS showed significant physical and chemical alteration when exposed to batch components. Therefore, its use in batch charging areas of the superstructure should be limited and carefully monitored. In glass contact application, such as feeder channels, vibro-cast AZS may show absorption of glass melt within its high level of porosity. The implications of this phenomenon during the furnace campaign need to be understood.

ALUMINA BASED REFRACTORIES
Table III shows physical and chemical properties of fused-cast and vibro-cast alumina refractories tested in this study. The vibro-cast alumina products are made up of alpha alumina grains bonded with cement; whereas fused-cast alumina in glass contact application is made up of both alpha and beta alumina grains, with a minor amount of nephelitic grain boundary phase. This difference can be readily seen in the typical chemistry of these products. Vibro-cast alumina products have more than 98% Al2O3 and very little Na2O; whereas fused-cast alpha beta alumina products have about 95% Al2O3, 3- 4% Na2O, and 0.5 – 1.0% SiO2. The higher Na2O in fused-cast products combines with a portion of the Al2O3 to form beta alumina phase.

The vibro-cast alumina product was found to have slightly lower thermal conductivity but a higher coefficient of thermal expansion than fused-cast alpha-beta alumina. The vibro-cast alumina microstructure consists of coarse-grained alpha alumina with a fine grained bond of alpha alumina and cement (See Figure 15). The porosity in the vibro-cast is much higher (18%) than in fused-cast alpha-beta alumina (<2%).

Table III: Key properties of Fused-cast alpha-beta alumina and vibro-cast alumina

Alumina Products for Glass Furnace Application Forehearth Channels & Covers; Refiner Paving							
Process	Al_2O_3	Major Phases	Bulk Density	Apparent Porosity	Thermal Conductivity W/m K		Thermal Expansion
	wt%		g/cc	%	1200C	1000C	% at 1200C
Fused-cast	94.5	alpha Alumina beta Alumina	3.17	< 2	3.7	3.74	0.88
Vibro-cast	98	alpha alumina	3	18	3.3	3.6	1.02

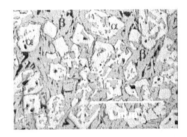 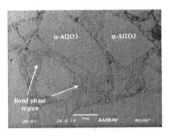

Figure 15: Reflected light micrographs of FC alpha-beta alumina (left) and vibro-cast alumina (right)

Whereas fused-cast alpha-beta alumina refractories have been used in the feeder channel application for many decades, the use vibro-cast alumina in feeder channels is more recent. In this application, the key performance criteria include glass melt contact corrosion resistance and glass defect potential. Service temperatures are expected to be in the range of 1250°C – 1350°C.

Corrosion from Contact with Glass Melt

As with AZS samples, we started our investigation with standard static finger corrosion test, using 12.5mm diameter core samples. First test was performed at 1300°C for 120 hours and the results are shown in Figure 16. The % area lost at the metal line was rather small, ~5.5% or less, and both fused-cast alpha beta and vibro-cast alumina samples showed similar corrosion resistance. In order to differentiate between these materials, the next test was performed at 1350°C for 120 hours. Following the test, all of the vibro-cast samples failed (see Figure 17), with the portion of the samples below the metal line separating from the rest of the fingers and dropping into the glass melt contained in platinum crucibles. In contrast, the fused-cast alpha beta alumina sample did not fail at the metal line and showed marginal area lost due to corrosion. Even significantly shorter test duration at 1350°C gave the same result.

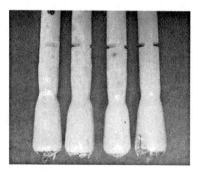

Figure 16: Corroded finger samples. The first three samples from left to right are different vibro-cast alumina samples, the sample on extreme right is fused-cast alpha-beta alumina

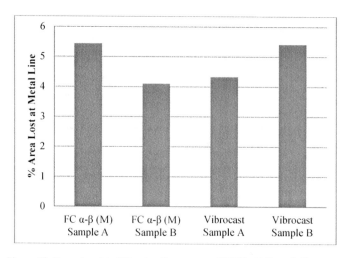

Figure 17: Corrosion data following finger test at 1300°C, 120h, soda lime glass

Due to the repeated failures of vibro-cast alumina samples in the standard finger corrosion test, a crucible test was performed at 1300°C for 96h. Following the test, all 3 vibro-cast alumina crucibles showed glass on the external surface of the crucible, whereas the alpha-beta alumina sample looked dry. Figure 18 shows the sawed cross-sections of the 4 crucibles. The three vibro-cast samples showed lower glass levels than the fused-cast sample, indicating *absorption* of glass within the porosity. Unfortunately, the drop in glass level during the test made any meaningful measure of the corrosion at the metal line impossible. Therefore, a modified finger corrosion test was developed.

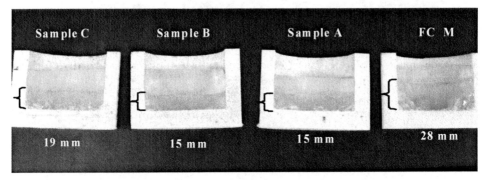

Figure 18: From left to right, 3 different vibro-cast alumina samples; sample on extreme right is fused-cast alpha-beta alumina. Note lower glass volume in all vibro-cast alumina crucibles.

The modified finger test used 22mm diameter core samples from the 3 vibro-cast and one alpha-beta alumina refractory placed in one large crucible, made out of fused-cast alpha-beta alumina (Figure 19). In comparison, the standard finger test uses 12.5mm diameter core sample placed individually in a platinum crucible. Therefore, the refractory to glass volume ratio is different in these tests. The modified test showed limited success. All the samples remained intact, but had little corrosion, after 24 hours, but two out of three vibro-cast fingers failed with additional 49 hours duration in the test; whereas the fused-cast sample did not fail (Figure 20).

Figure 19: Modified finger test. Left photo shows large crucible, right photo shows arrangement of samples

Figure 20: Modified finger test 1350°C, left photo shows samples after 24h, right photo shows samples after additional 49hours

At this point, it would be tempting to conclude that vibro-cast alumina refractory will not perform equal to the fusion-cast alpha beta alumina refractory. However, since vibro-cast alumina refractories are being used in feeder channel application, we still needed to understand the behavior of such materials in glass melt contact application. In particular, we wanted to further understand the *absorption* of glass melt within the porosity of vibro-cast alumina products. For example, does the absorption of glass melt continue with time? What is the effect of temperature cycling on refractory that is partially infiltrated with glass? Will it crack or spall? And so on. Therefore, a new test was designed to understand change in the vibro-cast material with thermal cycling.

This new test utilized vibro-cast crucibles placed tightly inside a fused-cast alpha-beta alumina holding crucible to minimize the leak of soda lime glass through the vibro-cast crucible. For a control sample, a fused-cast alpha-beta alumina crucible was also tested under the same conditions. First test was performed at 1350°C for 24hours. Following the test, the glass level was measured with a caliper from the top of the refractory crucible to the top of the glass. Then this test and the measurements were repeated five times. After 3 cycles, additional glass was added to the vibro-cast alumina crucibles due to rapid drop in glass level. This was not necessary for fused-cast alpha-beta alumina crucible.

The results of this temperature cycling test are shown in Figure 21. The vertical axis on this chart represents the distance of the glass surface from the top in inches. The glass in the fused-cast alumina crucible remained the same throughout the six test cycles. In contrast, the vibro-cast alumina crucibles showed higher glass distance immediately after first cycle indicating glass absorption. Further increase in the distance in cycles 2 and 3 show continuing absorption of glass melt within the porosity of the crucibles. The distance of glass changed from cycle 4 due to addition of fresh glass. However, cycles 5 and 6 show further absorption of glass, as noted by increasing distance of glass from the top of the crucibles. The results clearly show that, in an isothermal laboratory test, vibro-cast alumina refractories will absorb glass melt at operating temperatures. However, the rate of glass absorption slows down over time.

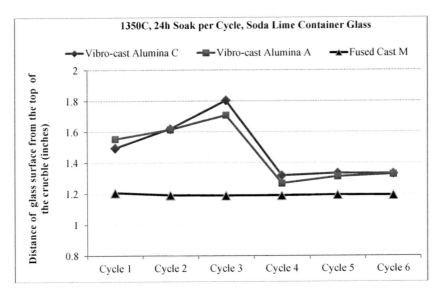

Figure 21: Thermal cycling test of vibro-cast alumina and fused-cast alpha-beta alumina crucibles

Glass Defect Potential

Figure 22 shows the alumina samples following the modified finger corrosion test described earlier. In a finger corrosion test, it is common to see a bead of glass adhering to the bottom of the samples. This bead of glass can be examined under a microscope to check for any refractory grains that may have become loose due to corrosion. Figure 22 shows a close-up of the glass bead at the bottom of one of the vibro-cast sample from the modified finger test. The glass bead clearly shows presence of a large refractory grain that may eventually become a stone in the glass product.

Figure 22: Large white colored stone observed

SUMMARY

The vibro-cast alumina refractories are coarse grained and porous products, containing mainly alpha alumina grains, in comparison to fused-cast alpha beta alumina refractory. Despite the large amount of porosity, the vibro-cast alumina products have marginally lower thermal conductivity than fused-cast alumina product. This would imply that heavy insulation would also be needed for vibro-cast alumina feeder channels. Both finger and crucible tests, used in this study, clearly showed superior performance of fused-cast alpha-beta alumina at 1350°C. It was also demonstrated that vibro-cast alumina products absorb glass in a short-term isothermal test in a laboratory furnace. This glass absorption increased quickly initially but slowed down over time in the cycling test. In a furnace application though, glass absorption may slow down or stop at some point in time due to temperature gradient within a feeder channel. This needs to investigated and understood.

CONCLUSIONS

Testing of coarse-grained and porous bonded refractories, such as pressed and sintered AZS and vibro-cast AZS and alumina, should be performed on large samples and for long durations (more than a few days), in order to have a measurable and reliable result. Fusion-cast refractories, both AZS and alpha-beta alumina, will generally perform better due to their low level of open porosity and tight interlocking crystals.

The use of bonded AZS refractories as paving blocks in glass furnace bottoms for the past several years suggests that these refractories can indeed be used, in place of fused-cast AZS, in some furnaces. However, this study clearly showed the inadequacy of common finger corrosion test and the use of below metal line data in accurate comparison of different bonded AZS refractories. Crucible test used in this study could not differentiate between different bonded AZS refractories available commercially.

The use of vibro-cast AZS in the batch charging areas of soda lime glass furnaces may lead to faster corrosion of these products and may be ill-advised in load bearing applications such as port arches, and for furnaces expecting full campaign life. In feeder channel application, higher rate of corrosion can be expected from vibro-cast AZS products, compared to fused-cast AZS.

The vibro-cast alumina products are also being currently used in some soda lime container furnaces in feeder channel application. The test performed in this study has clearly shown absorption of glass within the high porosity of vibro-cast alumina products. It can also be safely assumed that these products will experience a higher rate of corrosion than fused-cast alpha-beta alumina, though this study showed it is difficult to quantify the difference. The impact of glass absorption, higher rate of corrosion, coupled with potential for alumina stones needs to taken into account when selecting such products for feeder channel application.

Glass industry can benefit from thorough evaluation of post-campaign samples of bonded AZS pavers, vibro-cast AZS and alumina products. Such a study, coupled with glass defects data from furnace campaign, can be immensely useful in determining the cost vs. benefit of using lower-cost refractories.

ACKNOWLEDGEMENT
The authors are thankful for the cooperation of their colleagues at Technical Center, Leoben, Austria; Sales & Marketing in Wiesbaden; and Didier Werke Plant in Niederdollendorf, Germany.

ENERGY SAVINGS AND FURNACE DESIGN

Matthias Lindig-Nikolaus
Sorg GmbH, Germany

1. LEGISLATIVE CONDITIONS IN EUROPE

In April 2009 the European Union has brought to an issue the third emissions trading period. The procedure of granting emission certificates will change significantly. A general upper emission limit for the European Union was determined being valid from 2013 on. In total only 1.87 Billon tons of CO_2 will be allowed for the European Union. Every following year this limit will be reduced by 1,74%. In 2020 an overall CO_2 emission of 1,74 Billion tons should be achieved (about 20% less). This will be a challenge for the entire industry and in particular for the glass producing industry.

From 2005 to 2008 during the first trading period in Europe the CO_2 emissions were almost only monitored without any significant reductions. It was the introduction phase. The certificates for emissions were free of charge and the allocated amount was adjusted to the emissions which were generated during the previous time. A minor reduction by factor 0,9875 was requested.

During the second CO_2-trading period the European nations have committed themselves to reducing the total emissions by a certain percentage related to the years 1990 until 2012. The basic instrument for achieving this goal should have been the CO_2-trading market.
Even this commitment did not have a significant impact on operating conditions in the industry.

From 2013 on the emission limitation in Europe will be handled in a different way. The bench mark regulation will be implemented. For each produced ton of glass the producer receives a certain amount of CO_2 which equates the average emission of the 10% of the producers in Europe with the best performance (best available technology)! Those who are not among the 10% of best performance have to pay for extra CO_2-emissions. A CO_2 allocation list for Germany can be found in (1).

Furthermore the total emissions in Europe have to be reduced by 1,74% every year until the end of the third trading campaign in 2020. A so-called cross-sectorial correction factor will be used which regulates the national allocation of the CO_2-emissions. Details about this factor are not yet published. Decisions will be expected in the late fall this year regarding final national emission allocations.

What we have to anticipate is a shortage of at least 10% allocation for the majority of the producers. In the beginning of the CO_2-trading period costs of about 30€/ t of CO_2 were expected. Since more certificates were emitted than requested (about 5% in 2011) the cost for 1t of CO_2 went down 5-7€. This will certainly change during the next trading period.

2. IMPACT ON GLASS INDUSTRY
It is apparent that the environmental police in general do not consider the very individual production conditions of the glass industry and of course of many other industries. Glass production is characterized by long production facility lifetimes. Short term changes due to environmental

requirements cannot be attained. Aging of the facilities require increased specific energy demand until the end of the facility campaign. Very individual production conditions may demand to run the facility beyond the optimum operation conditions, i.e. the throughput has to be reduced due to the special item to be produced. Shortages in foreign cullet availability might also request increased amount of cullet.

What we can only look for, in order to fulfil the legal requirements, are in most of the cases subsequent measures for the existing facilities.

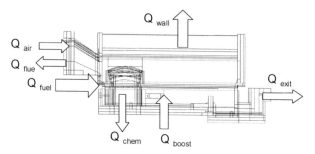

Fig.1 container glass endport fired furnace- energy balance variables

All possible options are described in figure 1 with the variables. Only these variables allow reducing the energy consumption and CO_2- emissions. The possibilities will be discussed in the following.

3. OPTIONS ON ENERGY SAVINGS AND CO_2 REDUCTION

The most basic option to reduce energy consumption will be an optimization of the batch composition. The options are to increase the share of alkali's, use of Calumite or addition of Fluorine. What is affected shown in figure 1 is the chemical heat demand and the heat content of the glass being released. About 2 wt% of Sodium Oxide increase allows to reduce the furnace peak temperature by almost 45°C. The fining temperature (dyn.viscosity 10exp2 dPas) is reduced by this temperature difference. In addition more alkali addition results in more silicate formation during batch to melt conversion. It helps to reduce the share of free silicate which has to be dissolved during the final melting process. A reduction in energy consumption by at least 3% will be possible. This equates to a CO2-reduction by about 2-3%. Calumite is affecting significantly the melting conditions. The batch agent is already in a glassy state and contains significant share of limestone, Alumina and Magnesia. In a case study 13% of addition results in an energy reduction by 4%. (2) Fluorine will also reduce at higher temperature the viscosity and will allow reducing the process temperature (3). 0,1-0,3% might be used without harming the manufacturing tools.
Improved batch recipe often can explain a good melting performance of a glass melting furnace (4). It should be carefully taken into account when bench mark studies are being done.

The wall losses are being reduced during the decades significantly by applying more and more effective insulation materials. In current state furnace design there is almost very little clearance for further improvements. An 250mm AZS wall in the superstructure use to be coated with about

130mm high Alumina Mullite (60% Al), 250mm Mullite (58%Al) and 150mm Insulation brick. This results in an almost uniform temperature within the AZS block. The risk of strong and continuous exudation of the glassy phase becomes highly potential.

Flue gas heat recovery is one of the major options we have looking at reduction in melter energy savings. The energy content of the flue gas behind the combustion air preheating is about 30% of the total energy input. The problems linked with heat transfer from flue gas to batch were discussed in previous papers from the author (5). The reaction of the batch ingredients with water is the key problem as well as the dusting outside and inside the furnace due to the hot and volatile batch.
The Sorg company has developed and installed a full scale batch preheating facility in October 2011 at Wiegand Glas Germany. The preheater is in operation since that time. A short and scheduled inspection was carried out in spring 2012. One of the features which helped to get over the key problem is the mechanical activation of the heat exchanger itself. The batch is fed from the top, the batch migrates due to gravity and continuous discharging below through the exchanger. The horizontal flue ducts can be activated in sections and prevent clumping or caking of the material. The total device is self cleaning. In order to prevent dusting out side and inside the furnace the Sorg company has invented an enlarged doghouse with a charging system which allows delivery of the batch within an air dense system, which allows separation of the batch into thin and small clumps and which allows glazing of the clump surface. By glazing the clumps surface batch dusting in the furnace will be reduced significantly. The principles of this solution were presented previously by the author (5). The operation figures and additional measuring results are given in the following table.

Table 1: container glass furnace without and with batch preheating. CO_2 reduction with preheating about 13%

		Feb.2011 no preheat	April 2012 batch preheat
throughput	t/24h	246	256
cullet	%	70	69,5
boosting	kW	769	410
nat.gas	Nm³/h	1076	1012
batch temperature	°C	20	220-240
batch moisture	%	2 - 3	2-3
spec.energy	kJ/kg	4082	3587
savings	%		12,13
CO2 from Combustion	kg/t	210	190
CO2 from Boosting	kg/t	32	16
CO2 from Batch	kg/t	39	38
CO2 total	kg/t	281	244

The implementation of the heat exchanger was done after starting the furnace operation. The glass color was changed several times without any impact on furnace stability. Even very redox- sensitive colors like amber and dead leaf were successfully produced. The expected reduction in energy consumption was achieved. The dusting inside the furnace was checked by measurements. The ambient around the furnace was completely dust free. The facility is air dense.
Total dust was measured on top of the regenerators and behind the regenerators. Comparative measurements were carried out at other furnace with same size and running under similar conditions

except batch preheat. The amount of dust in the flue gases did not differ! The result can be simply checked by inspection of the checker packing. No plugging of the packing was observed after 8 month of operation. The principle of the special charging solution was proven well-founded.

The flue gas behind the preheater exits with a temperature of about 240°C. Minor improvements in the preheater itself might allow decreasing the flue gas temperature down to 190°C. It will depend on customers boundary conditions. The flue gas has to be cleaned by a precipitator which requires a minimum temperature of about 230°C. From this point of view it is more reasonable to add as appropriate a power- heat coupling. It will be possible to run behind the precipitator a ORC processing with an efficiency of 10-20%. This measure will not help indirectly to reduce CO_2 emissions.

With batch preheat operation it might also be recommended to reduce application of electric boosting as much as possible. E-boost application might reduce CO_2-emission on short term relevant to the facility only. Electric power generation produces about 0,214m3/kWh (0,421kg/kWh) CO_2. A comparative analysis of the CO_2- Emission for different operation conditions is given in the following table

Table 2: comparative analysis influence of batch preheat and boosting on CO_2- emission

Endport		
melting area	m^2	90
throughput	t/d	260
cullet	%	70
color		green

		no BPH low boost	no BPH high boost	BPH low boost	BPH high boost
nat gas	m^3/h	1167	1098	1018	957
boost	kW	300	800	300	800
batch	°C	20	20	227	208
spec.energy	kWh/t	1,11	1,09	0,97	0,96
CO_2 total	kg/t	258	265	231	239
CO_2	%	2,5		3,5	
$CO_{2\ flue}$	kg/t	216	203	189	177
$CO_{2\ boost}$	kg/t	12	31	12	31
$CO_{2\ batch}$	kg/t	30	30	30	30

4. SUMMARY

The spirit of the Kyoto -protocol is going to become an issue for the global industry. The European nations have established a program which is constituted for long term. National legal demands are following. The contribution of the glass industry to the overall energy saving and CO_2- emission reduction program is limited by essential physical limitations. We will have to anticipate a legal demand for CO_2- reduction in Europe from 2013 onwards. There is no time for long term investigations in fundamental new ways of melting. Batch composition optimization and batch preheat are the two essential options to enable the industry to be inline with the future legal demands. New technologies, Carbon-free combustion might be something which has to be investigated next and with strong effort.

REFERENCES

(1) European allocation list available under: http:www.dehst.de
(2) N.J.Meriott, Reduce furnace energy consumption using Calumite, Calumite Ltd.UK, company information
(3) Götz, the effect of fluorides on melting, 10th Int.Congress on Glass Kyoto 1974
(4) R.Conradt, reaction kinetics of particulate matter, 11th ESG Conference, Maastricht, 2012
(5) M.Lindig, Initial experience with Sorg integrated batch handling concept, 16th ATIV conference, Parma, 2011

NEW TIN OXIDE ELECTRODES FOR GLASS MELTING

Julien Fourcade and Olivier Citti
Saint-Gobain SEFPRO – Northboro R&D Center
9 Goddard Rd, Northborough, MA, USA

ABSTRACT

Tin oxide is a refractory and a semi-conductor which make it an excellent candidate for electrodes in electric glass melting. Tin oxide electrodes have been used for many years for the melting of Crystal glass, borosilicate glasses such as LCD and other types of glasses. It can be used as a sole source of power in electric furnaces or as electrical boosting with gas/fuel fired furnaces. Tin oxide electrodes can be found in various shapes (cylinders, bullets, rectangles, and special shapes) and sizes from a few inches long up to 25 inches.

Tin oxide is poorly soluble in silicate melts and it is often presented as the second most corrosion resistant refractory after chromium oxide for most silicate melts. At high temperature, however, stannic oxide (SnO_2) tends to be reduced into the more soluble and more volatile stannous oxide (SnO) according with the following reaction: $SnO_2 = SnO + \frac{1}{2} O_2$

Reduction of tin oxide, therefore, enhances corrosion by the melt and decreases significantly the lifetime of electrodes.

The authors have developed and patented new tin oxide materials for electrodes with enhanced stability at high temperature to improve corrosion resistance in silicate melts and in air. New electrodes were designed to have extended lifetime without affecting their electrical properties.

Keywords: Electrical Melting, Tin oxide Electrode, Refractories, Corrosion.

INTRODUCTION

Tin oxide is typically found as tin monoxide (SnO) also called stannous oxide and as tin dioxide (SnO_2) or stannic oxide. As shown in Figure 1, tin monoxide (stannous oxide) adopts a tetragonal structure and tin dioxide (stannic oxide) has rutile structure (hexagonal structure). Stannic oxide can be naturally found as cassiterite.

Tetragonal structure   Hexagonal structure

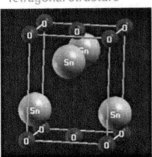 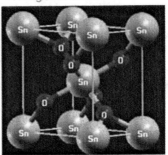

Figure 1. Image of the crystallographic structures of tin monoxide and tin dioxide

Stannous oxide is used as anode material in rechargeable Lithium-ion batteries, as coating materials or as catalysts. SnO(s) is metastable at room temperature and decomposes at about 600K upon heating: SnO(s) → SnO$_2$(s) + Sn(l). Stannic oxide is used as reflecting foils, transparent electrodes in liquid crystal displays (LCDs), chemical gas sensors and Sn-oxide/Si solar cells, and electrodes for electric glass melting furnaces. SnO$_2$(s) is stable in air and reduces at temperature above about 1300K: SnO$_2$(s) → SnO(g) + ½ O$_2$(g).

Tin oxide electrodes have been used to melt glass for more than 50 years. In October 1965, Plumat *et al.* (Plumat, Jaupain, & Toussaint, 1965) presented a paper on the use of tin oxide bricks as glass melting electrodes at a glass conference.

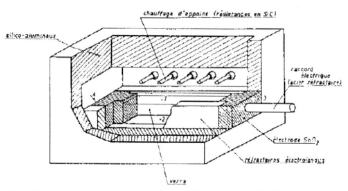

Fig. 15a - Bassin expérimental avec électrodes en SnO$_2$.

Figure 2. Schematic view of the experimental setup described by Plumat et al.; on the drawing fused cast are used as sidewalls for glass contact and tin oxide bricks are placed at each end of the tank. Tin oxide bricks are connected with refractory metal rods for power. The setup is placed in silica-alumina casing and additional heat is provided above glass with SiC heating elements.

Tin dioxide (SnO$_2$) has the second lowest solubility, after chromium sesquioxide (Cr$_2$O$_3$) in most silicate melts. SnO$_2$ moreover has very good glass contact qualities: very low stoning potential and low coloring potential. Finally, SnO$_2$ is a n-type semiconductor with a band gap of 3.6 eV; which makes it a very attractive electrode material for electrical glass melting or electrical boosting in many silicate glass melts.

Standard commercial electrodes are manufactured by Corhart Refractories (Saint-Gobain SEPR) under the name T-1186. These electrodes are doped with copper oxide and antimony oxide to allow sintering up to about 93% of theoretical density and to have good electrical conduction from room temperature up to operating temperature. Typical properties for T-1186 electrodes are reported in Table 1 and a view of the microstructure is presented in Figure 3.

Tin oxide electrodes can be manufactured in various sizes and shapes and they can be used as bottom electrodes or as sidewalls electrodes. Tin oxide electrodes must be totally immersed to prevent rapid deterioration throughout sublimation of SnO.

Table 1: typical properties of T-1186 electrodes manufactured by Corhart Refractories (source: SEFPRO catalogue)

Typical chemical analysis	
SnO_2	98.5
CuO	1.5
Sb_2O_3	
Typical physical properties	
Density (g/cc)	6.5
Apparent Porosity (%)	5
Cold MOR (MPa)	45
Electrical Resistivity (Ω.cm)	
100°C	0.0100
400°C	0.0100
1000°C	0.0015
1400°C	0.0010

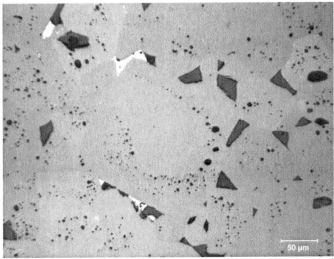

Figure 3. Microstructure of T-1186, observed with optical microscope after 1 micron polishing of a cross-section. Grey phases are the tin oxide grains, white spots are copper-rich phases that precipitated at grain boundaries, and dark grey areas are porosity.

CORROSION OF TIN OXIDE REFRACTORY

Dynamic corrosion tests (Merry-Go-Round) were performed with 2 tin oxide samples (Ø=20mm) in LCD glass at different temperature from 1550 to 1650°C for 24 to 90 hours. Time was adjusted with temperature and corrosion to avoid saturation of the glass. As reported in Figure 4, corrosion of tin oxide refractory in LCD glass melt follows an Arrhenius law in the domain 1550 - 1650°C: $k = A.e^{-Ea/RT}$ and activation energy 726 kJ was measured.

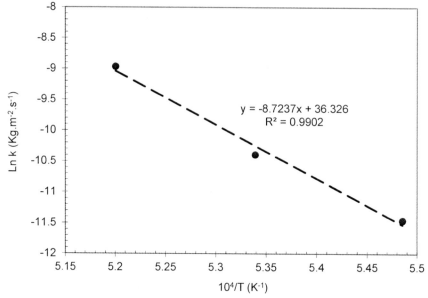

$$y = -8.7237x + 36.326$$
$$R^2 = 0.9902$$

Figure 4. Corrosion rate of tin oxide in kilograms per surface and time unit in LCD glass and plotted in function of $10^4/T$.

As previously reported by Benne et al. (Benne & Russel, 2005) and by Farges et al. (Farges, Linnen, & Brown, 2006) SnO_2 dissolves directly into glass melts and $Sn(IV)$ occupies similar sites as $Zr(IV)$ in most melts. Therefore, $Sn(IV)$ is surrounded by 6 oxygens with 4 bridging oxygen (BO) and 2 non-bridging oxygen (NBO). For charge compensation, $[SnO_6]^{2-}$ units need anions such as Na^+ or Ca^{2+}; therefore solubility of $Sn(IV)$ is increased in melts that contain NBO; hence melts with alkali and alkali-earth over aluminum ratio greater than 1. Also, according to Farges et al. (Farges, Linnen, & Brown, 2006) and Linnen et al. (Linnen, Pichavant, & Holtz, 1996) SnO_2 can dissolve into the melt as $Sn(II)$ and $Sn(II)$ is incorporated into the melt as a more ionic specie, similar to Ca^{2+} and Pb^{2+}. For this reason, $Sn(II)$ is predominant in melts with low contents of NBO; and its mobility is believed to be higher than $Sn(IV)$.

Solubility of tin species (Sn^{4+} and Sn^{2+}) were calculated by Condolf (Condolf, 2008), using FactSage in an alkali-free glass melt as function of temperature and pO_2. Composition of the glass melt is reported in Table 2. Results for solubility of tin in temperature from Condolf (Condolf, 2008) are shown in Figure 5 for various pO_2. First, solubility of tin in LCD melt increases with temperature, then above 1300°C, solubility of tin increases further as pO_2 decreases. At 1550°C for example, solubility of tin in LCD melt is about 2.8% (weight percent) in oxidized melt (pO_2=1atm) and solubility increases up to 3.9% in more reduced melt with pO_2=0.1atm. Note that similar solubility is achieved at 1650°C in oxidized melt (pO_2=1atm). As shown in Figure 6, Sn^{2+} content is increasing as temperature increases and oxygen partial pressure decreases. Based on Figure 5 and Figure 6 at T=1550°C and pO_2=1atm, solubility of Sn^{4+} is 2.2% and solubility of Sn^{2+} is 0.6%. When oxygen partial pressure decreases to pO_2=0.1atm, solubility of Sn^{4+} only slightly increases to 2.3% whereas solubility of Sn^{2+} increases to 1.6%.

Table 2: composition of the alkali-free glass melt used for calculation using FactSage

	Weight %
Al_2O_3	14.44
B_2O_3	9.76
CaO	11.19
Fe_2O_3	0.02
Na_2O	0.20
SiO_2	63.87

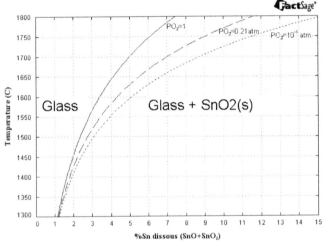

Figure 5. Solubility curves of tin in LCD glass melt in temperature for SnO and SnO_2 combined plotted for oxygen partial pressure of 1, 0.21 and 0.1 atm.

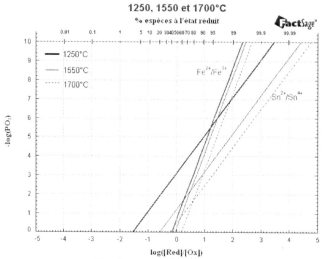

Figure 6. Redox couples (Fe^{2+}/Fe^{3+}) and (Sn^{2+}/Sn^{4+}) calculated as function of oxygen partial pressure in the melt and plotted at 1250, 1550 and 1700°C.

Based on these results, SnO_2 is believed to dissolve into the melt as both Sn(IV) and Sn(II). Sn(II) plays an increasing part in the process as temperature increases and pO_2 decreases, especially in AF glass melts. Tin oxide with superior resistance to reduction at the interface with the melt should have enhanced corrosion resistance at high temperature. New Zr-doped tin oxide electrodes were developed to enhance corrosion resistance: T-1186 ZR and T-1188 ADV.

NEW TIN OXIDE ELECTRODES

Electrode properties for T-1186 Zr and T-1188 ADV are listed in Table 3 and microstructures are shown in Figure 7 and Figure 8. T-1186 ZR has very similar properties and microstructure as T-1186; Zr^{4+} is in solid solution within SnO_2 grains and therefore it is not visible in the microstructure. On the other hand, T-1188 ADV is quite different from T-1186; density is slightly higher and apparent porosity is lower than 1%. Microstructure shows finer grains well bonded and essentially no precipitate at grain boundaries. Electrical resistivity is still very low from room temperature up to operating temperature. The difference in microstructure is due to the change in sintering aid with the introduction of zinc oxide to replace copper oxide.

Table 3: typical properties of new tin oxide electrodes

	T-1186	T-1186 ZR	T-1188 ADV (advanceable)
		Patent Pending	Patented
Typical chemical analysis			
SnO_2	98.5	96.5	96.5
CuO / ZnO	1.5	1.5	1.5
Sb_2O_3			
ZrO_2	-	2	2
Typical physical properties			
Density (g/cm^3)	6.5	6.6	6.6
Apparent Porosity (%)	5	2.5	0.5
Electrical Resistivity (Ω.cm)			
100°C	0.01	0.01	0.02
400°C	0.01	0.01	0.02
1000°C	0.0015	0.002	0.003
1400°C	0.001	0.002	0.003

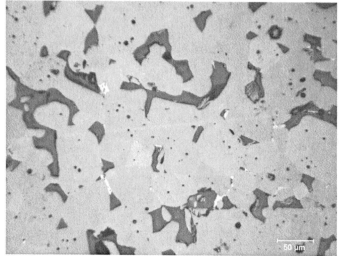

Figure 7. Microstructure of T-1186 ZR, observed with optical microscope after 1 micron polishing of a cross-section. Grey phases are the tin oxide grains, white spots are copper-rich phases that precipitated at grain boundaries, and dark grey areas are porosity. ZrO_2 is not visible in the microstructure as Zr^{4+} is in solid solution within SnO_2 grains.

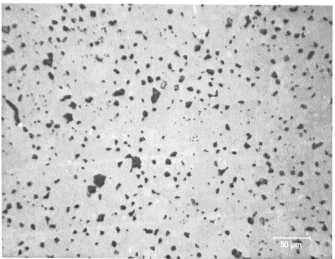

Figure 8. Microstructure of T-1188 ADV, observed with optical microscope after 1 micron polishing of a cross-section. Grey phases are the tin oxide grains and dark grey areas are porosity. Copper-rich phases that precipitated at grain boundaries are essentially absent from

microstructure and ZrO_2 is not visible in the microstructure as Zr^{4+} is in solid solution within SnO_2 grains.

Observation of a piece of tin oxide electrodes after several months in LCD glass, in Figure 9, shows cracks several centimeters away from interface. Microstructure observations in several areas of the electrode, noted "A", "B", "C" and "D", are shown in Figure 10 to Figure 13. Areas "A" and "C" present the most changes from initial microstructure. In the hot zone, noted area "A", grains and pores grew significantly and pores got oriented along temperature gradient. Also, copper phases are totally depleted in the hot zone. In the cracked zone of the electrode, noted area "C", we observed large accumulation of copper oxide precipitates at grain boundaries. We suspect these phases to create thermo-mechanical stresses that are responsible for cracking. Copper oxide phases become liquid at high temperature (T>1200°C) and tend to migrate away from the hot zone to precipitate at a temperature near 1100-1200°C. Based on the Cu-O phase diagram, this is the domain of liquidus for both Cu_2O and CuO.

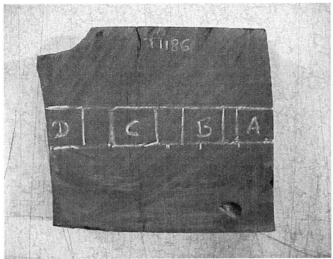

Figure 9. Picture of a piece of T-1186 electrode after several month in LCD glass melt; the glass was on the right hand side on the picture. Samples were prepared for polishing in areas noted "A", "B", "C", and "D".

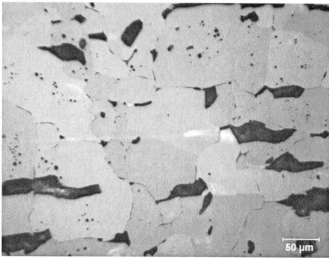

Figure 10. Microstructure of T-1186 in area "A" observed with optical microscope after polishing. Grains are large (>50 μm) and pores are large and elongated. Copper-rich phases have essentially disappeared from microstructure.

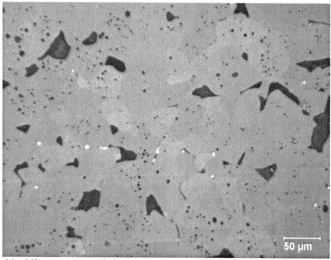

Figure 11. Microstructure of T-1186 in area "B" observed with optical microscope after polishing. Some grains and pores are larger but orientation has disappeared and copper-rich phases are observed at grain boundaries.

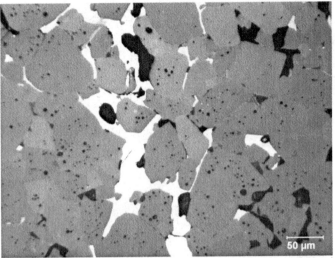

Figure 12. Microstructure of T-1186 in area "C" observed with optical microscope after polishing. There is no major change in the grains or pores but large quantities of copper-rich phases have precipitated at grain boundaries. These larges precipitated phases are believed to induce cracking observed on the electrode in Figure 9.

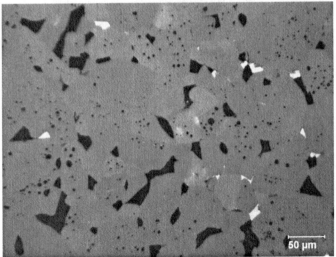

Figure 13. Microstructure of T-1186 in area "D" observed with optical microscope after polishing. There is no change from initial microstructure.

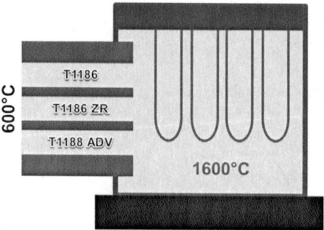

Figure 14. Schematic view of the copper migration re-heat test developed in Saint-Gobain CREE. Tin oxide samples are placed in the door of an electric box furnace for 100 hours to be re-heated under thermal gradient. The hot face was measured at 1600°C and the cold face was measured at 600°C.

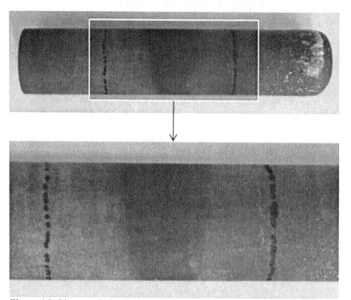

Figure 15. Picture of the T-1186 electrode after 100 hours under thermal gradient in the copper migration re-heat test. Accumulation of copper oxide is clearly visible in red at the center of the electrode, in an area at about 1200°C.

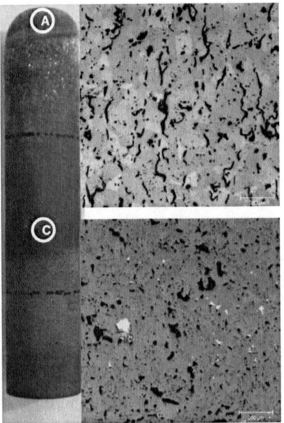

Figure 16. Picture of the T-1186 ZR electrode after 100 hours under thermal gradient in the copper migration re-heat test and microstructures observed in areas "A" and "C" of the electrodes after sampling and polishing. Significant pore growth is observed in the hot zone "A" with orientation along thermal gradient. The hot zone is depleted of copper phases at grain boundaries that accumulate in area "C".

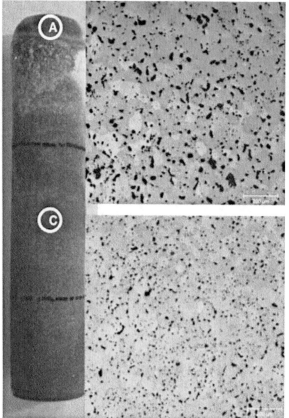

Figure 17. Picture of the T-1188 ADV electrode after 100 hours under thermal gradient in the copper migration re-heat test and microstructures observed in areas "A" and "C" of the electrodes after sampling and polishing. Some pore growth is observed in the hot zone "A" but without orientation and no copper precipitated in area "C".

Dynamic corrosion and power corrosion tests were performed in LCD glass at 1620°C. Dynamic corrosion results, in Figure 18, show about 40% lower sublimation for T-1186 ZR and T-1188 ADV than for T-1186. The results on corrosion by the melt are less significant; yet T-1186 ZR and T-1188 ADV present 15% lower corrosion than T-1186.

Power corrosion test setup is described in Figure 19. After testing of T-1186 and T-1186 ZR in the same test at 1620°C for 48 hours with 1A of alternative current, 34% (in volume) corrosion was measured on T-1186 and only 18% on T-1186 ZR.

Sublimation (Kg/m²/s)   4.4x10⁻⁵   2.6x10⁻⁵   2.4x10⁻⁵

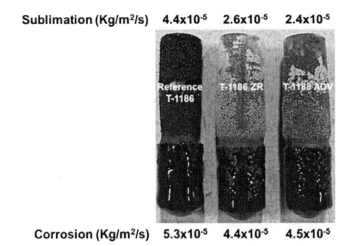

Corrosion (Kg/m²/s)   5.3x10⁻⁵   4.4x10⁻⁵   4.5x10⁻⁵

Figure 18. Results of dynamic corrosion at 1620°C for 48 hours in LCD glass. Sublimation of T-1186 Zr and T-1188 ADV is about 40% lower than for T-1186 while corrosion is about 15% lower.

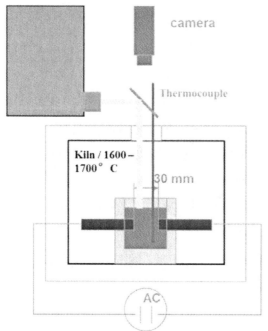

Figure 19. Schematic view of power corrosion test developed in Saint-Gobain CREE. Glass cullets are melted inside a zircon crucible. The crucible is pierced to allow positioning of tin oxide electrodes. Electrodes are connected to a power source with Pt wires and the whole setup is placed in an electric box furnace. The top of the crucible is open to allow refill of glass cullets and observation from the top of the furnace. Temperature of the glass is controlled with a thermocouple.

CONCLUSIONS

Tin oxide has high corrosion resistance in most silicate melts up to high temperature, low coloring potential, good electrical conductivity. Although it can be used as refractory, tin oxide electrodes are successfully used for electrical boosting or electrical melting of several glasses: Lead Crystal Glass, Alkali-free boro-silicate (LCD, OLED), Alkali rich alumino-silicate, ... Tin oxide electrodes can be manufactured in various shapes (cylinders, rectangles) and sizes. Tin oxide electrodes can be used vertically as bottom electrodes or horizontally as sidewall electrodes. Tin oxide electrodes are compatible with numerous types of refractories for electrode blocks: alumina, zircon, HFCZ, ... New Tin Oxide Electrodes (T-1186 ZR and T-1188 ADV) showed enhanced sublimation and corrosion resistance in AF melts. Moreover, T-1188 ADV showed superior stability of its microstructure in the hot zone of the electrodes and no accumulation of sintering aids at grain boundaries.

BIBLIOGRAPHY

Benne, D., & Russel, C. (2005). Diffusivity and incorporation of tin in Na2O/CaO/Al2O3/SiO2 melts. *Journal of Non-Crystalline Solids 351*, 1283-1288.

Condolf, C. (2008). *Influence d'un dopage ZrO2 sur la stabilité de SnO2 au contact du verre LCD : approche thermodynamique.* Aubervilliers: Saint-Gobain Recherche.

Farges, F., Linnen, R. L., & Brown, G. E. (2006). Redox and speciation of tin in hydrous silicate glasses: a comparison with Nb, Ta, Mo and W. *The Canadian Mineralogist 44*, 795-810.

Linnen, R. L., Pichavant, M., & Holtz, F. (1996). The Combined Effect of fO2 and Melt Composition on SnO2 Solubility and Tin Diffusivity in Haplogranitic Melts. *Geochimica et Cosmochimica Acta, Vol. 60, No 24*, 4965-4976.

Plumat, E., Jaupain, M., & Toussaint, F. (1965). Les Briques d'Oxide d'Etain et leur Utilisation comme Electrodes de Fusion. *Giornate del Vetro*, (pp. 5-12). Rome.

ACKNOWLEDGEMENTS

Authors would like to thank people who participated actively to testing and modeling in the course of that project and have contributed to some results presented in this document: Cyril Condolf from Saint-Gobain Recherche, Derek Payne from Corhart Refractories, Patrick Boistel and Dorothee Vernat from Saint-Gobain CREE.

STRUCTURAL HEALTH MONITORING OF FURNACE WALLS

Eric K. Walton[1], Yakup Bayram[1], Alexander C. Ruege[1], Jonathan Young[2], Robert Burkholder[2], Gokhan Mumcu[3], Elmer Sperry[4], Dan Cetnar[4], Thomas Dankert[5]

[1]PaneraTech, Inc., Columbus, OH
[2] The Ohio State University, Columbus, OH
[3]University of South Florida, Tampa, FL
[4] Libbey Glass, Toledo, OH
[5]Owens-Illinois, Perrysburg, OH

ABSTRACT

Erosion of the refractory lining in molten glass furnaces is a major problem for the glass manufacturing industry. When erosion on the walls is not detected early enough, it may lead to molten glass leaking through the refractory lining and resulting in suspension of production for several weeks and in some cases resulting in catastrophic accidents. Accordingly, glass manufacturers have to shut down their furnaces based on a conservative schedule to avoid any catastrophic molten glass leakage. Currently, there is no technology that can deterministically measure erosion of the furnace walls.

Under a National Science Foundation Small Business Innovative Research Program, PaneraTech, in collaboration with The Ohio State University, University of South Florida, Libbey Glass and Owens-Illinois, has recently demonstrated feasibility of a non-destructive wireless sensor technology for 3-Dimensional imaging of refractory lining in furnace walls to deterministically identify wall erosion. We demonstrated that we could see through AZS refractories with in-house measurements including 3 inch and 6 inch AZS blocks heated up to operational temperatures and also with in-situ measurements at an operational furnace at Libbey glass facilities. We also demonstrated that we can image fine features at the AZS–molten glass interface such as a groove resulting from flow of the molten glass. Lastly, the same wireless system has been shown to be capable of detecting voids and defects in cold refractory bricks before installation.

In this paper, we will discuss the underlying fundamentals of the proposed wireless sensor technology, the measurement results pertaining to feasibility and in-situ tests and the path-forward to an integrated wireless sensor system for structural health monitoring of furnace walls.

INTRODUCTION

Erosion of refractory lining of glass furnace walls is a major problem for the glass manufacturing industry. At some point, a glass furnace needs to be shut down and the walls replaced. Errors in the estimation of the thickness of the glass can lead to expensive replacement of furnaces with life still left or in some cases, to catastrophic failure of the walls with associated danger to human life, capital equipment around the furnace and significant production disruption. This paper discusses a research supported by a National Science Foundation Small Business Innovative Research Program for the development of a microwave probe system that can measure the thickness of such walls from the outside of the furnace. We will discuss the underlying fundamentals of the wireless sensor technology, the fundamental measurement tests and the path-forward to an integrated wireless sensor system for structural health monitoring of furnace walls.

COLD AZS STUDIES

We began this research by measuring the radio frequency (RF) transmission properties of AZS refractory material at room temperature. This refractory material is very commonly used to make the walls of glass manufacturing furnaces. We set up a fundamental test using laboratory microwave instrumentation to characterize the internal propagation characteristics of AZS over a very wide frequency band so that we could eventually find the optimum frequency band for probing. We measured the direct (through the material) propagation characteristics as well as the internal reflection characteristics. We were able to determine the room temperature propagation characteristics of attenuation versus distance as well as propagation velocity associated with the material (technically, we measured the complex dielectric constant of the AZS material as a function of frequency). An example of an experimental setup to measure the internal reflections of multiple layers of refractory material is shown in Figure 1. In this setup, a laboratory network analyzer is used with ultra wide band antennas to probe the material. Measurements were made over the entire frequency band from 2 to 18 GHz in this case. Then, the radar data was transformed to the time domain so that internal material reflections as a function of propagation time could be plotted. The plot in Figure 1 shows internal reflections. Note that distinct reflection terms can be identified for each discontinuity in the AZS material. It is possible to see internal gaps (associated with the stack of blocks) as well as a metal plate on the table beneath the blocks. The room temperature AZS material is transparent enough at room temperature so that accurate internal structures can be clearly characterized over these microwave frequencies. We know from the literature, however, that the attenuation of such signals will increase dramatically at the higher temperatures found in the glass furnace.

Experiments were done with several thicknesses of AZS blocks as well as with an AZS block with voids (drilled holes). It was shown that at room temperature, it is possible to accurately measure the thickness of the blocks and to map internal voids.

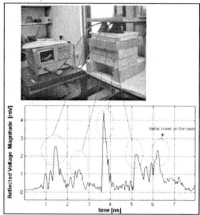

Figure 1. Experimental setup (top) and reflected microwave signal as a function of time (bottom) from a room temperature probing experiment.

HOT AZS STUDIES

It is known that the attenuation and the dielectric constant of AZS material increases as the temperature of the AZS increases, so it was necessary to make similar fundamental measurements of hot AZS. This was done using a small kiln. The top cover of the kiln was removed and replaced with blocks of AZS to be tested at high temperature. Thermocouple probes were placed at the bottom of the test blocks and at the top of the test blocks. Insulation material covered the blocks except where probe antennas were mounted. A sketch of the kiln setup is shown in Figure 2.

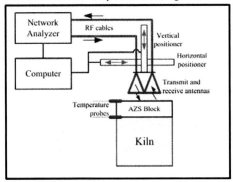

Figure 2. Diagram of the kiln setup showing the location of the AZS blocks and the temperature probes. Note the computer data collection system and antenna positioning system.

A large set of data on propagation of microwaves in hot AZS was collected using this system. An example attenuation plot is shown in Figure 3.

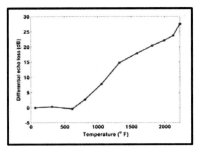

Figure 3. Microwave excess reflected signal loss in a 6 inch thick block of AZS as a function of temperature.

Figure 3 shows the microwave excess signal loss through 6 inch thick AZS for a reflection from the inner surface. Note the 28 dB RF loss (round trip reflection) due to the high temperature at

2300° F. This is in addition to the normal room temperature RF attenuation. In other words, one would expect a total of approximately 68 dB round-trip RF attenuation through 6 inch thick AZS at 2300° F (40 dB at room temperature and 28 dB additional due to the high temp). However, we must note Figure 3 displays the composite loss. In other words, the temperature through the thickness of the 6 inch block has a gradient as might be expected in an operational glass furnace. The top surface was approximately 1100° F, while the inner surface (bottom) of the AZS was at 2300° F. (It does not represent the excess loss through a 6 inch block with uniform 2300° F temperature.)

Of particular interest is the behavior of microwave internal reflections data associated with non-uniform perturbations in the inner walls of the glass furnace. This was tested using a 3 inch thick block of AZS with a 1 inch radius groove in the bottom. A scan at high temperatures over the outside surface was done, and the results processed to yield a type of microwave image of the inner surface of the block. An example image is shown in Figure 4.

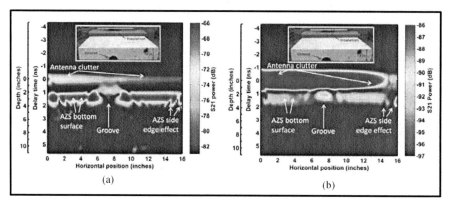

Figure 4. Microwave image of groove in bottom of 3 inch block of AZS. (a) 505 F; (b) 2200 F.

Note from Figure 4 that the reflections from the bottom of the block and from the groove are greatly attenuated at the higher temperatures, but still identifiable in the results.

ANTENNA DESIGN

The experimental testing of the AZS blocks also confirmed that the laboratory antennas that we were using were not efficient for probing the AZS blocks. These laboratory ultra wide band antennas are designed to radiate signals into the air, but in our case, we need to radiate microwave signals into the AZS block. Our tests confirmed that the dielectric constant of the AZS material has a value near 9. Thus, we need to develop antennas that are matched to a dielectric constant of 9. It is also clear that the antennas must not have internal reflections. Internal reflections in the antenna can confuse the interpretation of the microwave reflection profiles because they generate spurious reflection terms in the data.

To overcome this problem, antennas have been designed with wide bandwidths that are matched to the AZS dielectric constant of 9. An example of such a theoretical design is shown in Figure 5. In this case, the antenna is a spiral traveling wave design with a tapered balun printed on sheets of high temperature and high dielectric ceramic. Note the reflection coefficient plot (right hand plot in Figure 5). It shows that the antenna is matched much better to the AZS material (red line) than to air (black line). We built a right hand circularly polarized antenna for transmit and a left hand polarized antenna for receive to enhance the reflection terms from the inner face.

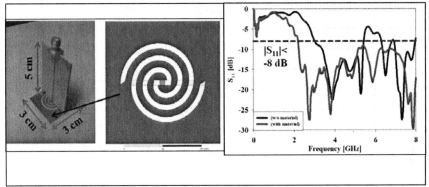

Figure 5. Design (left) and performance (right) of spiral high temperature antenna for AZS probing.

TEST RESULTS

A set of scanning results from the kiln tests is shown in Figure 6. In this set of experiments, the pair of spiral antenna was scanned across the top face of the 6 inch thick block as the temperature of the kiln slowly increased. At each scan position, a time domain transformation was performed. The set of scans is plotted as a color image showing reflected signal strength versus time delay for each position. The color represents the strength of the reflection term. It can be seen that the reflected term from the inner surface of the 6 inch block becomes much weaker with respect to the spurious coupling terms as the temperature increases. (The color scale is adjusted as the temperature increases.)

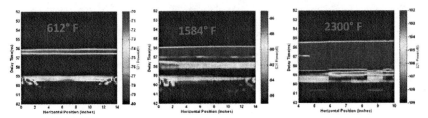

Figure 6. Color images of data from the microwave scanning versus temperature test sequence.

CONCLUSIONS AND FUTURE WORK

At this stage of the research, it is clear that we have developed a very sensitive technique that can permit us to see the inner surface of the refractory wall for the case of high temperature AZS. As the temperature increases, the material becomes very lossy. We have shown that properly designed radar-type probing and signal processing can overcome this loss. The remaining problem is to suppress the spurious internal reflections that can confuse the user as to where the actual reflection terms due to the inner surface are.

Finally, the sensor antennas and the microwave system must be packaged into a unit that the industrial user can easily set up and use. The signal processing must be converted to a semi-automatic user friendly package that can be used on a day to day basis by industrial technicians in the actual furnace environment. Data must be collected over the lifetime of the glass furnace so that the trends can be used to predict actual time to failure. Localized "thin spots" need to be detected and characterized.

In the future, we plan to develop an imaging system. This system will create a 3-D image of the inner wall of the glass furnace. Actual images of the inner surface of the walls will permit detailed analysis of problem areas and the development of mitigation techniques. Again, this imaging system needs to be commercialized to the point where trained technicians can use the system and provide images to the glass company engineering team.

NEW CIP SEFPRO REFRACTORY SOLUTION TO EXTEND SODA LIME GLASS FURNACE LIFE

C. Linnot[1], M. Gaubil [2], T. Consales, [2], O. Citti [3], T. Champion [2], J. Poiret [4]

[1] Saint-Gobain SEFPRO, Vénissieux, France
[2] Saint-Gobain CREE, Cavaillon, France
[3] Saint-Gobain NRDC, Northboro USA
[4] Saint-Gobain SEFPRO, Le Pontet, France

ABSTRACT

SEFPRO will present in this paper new Cast In Place "CIP" refractory solutions for cold repairs in soda lime glass furnaces. These new materials, based on fused cast AZS and corundum grain have been developed to increase the campaign life of a glass furnace. These new solutions for glass contact and superstructure materials will help glassmakers to reach their targets in terms of furnace safety and glass quality during extended campaign runs. Well adapted to the CIP process, these self leveling and pumpable products exhibit particular good thermal stability and glass contact properties. Lab and industrial results will be discussed during the presentation.

SEFPRO CIP SOLUTIONS

Glass makers are looking for some new solutions to repair their furnaces. The Cast in Place solution has already been frequently chosen on the insulation market for urgent or planned repairs. This solution exhibits several advantages:
- Reactivity and fast installation process
- Reduced amount of refractories to be ordered
- No risk of fragile block breakage

Most of the furnaces are candidates for the cast in place process, providing it is drained successfully. The weak parts of the blocks are removed prior to the casting in order to ensure a good mechanical strength of the new concrete.

Thanks to a Cast In Place process, the dedicated concrete is mixed and then poured or injected in some special forms on the top of the corroded refractory blocks to create a new shape.

SEFPRO provides the concrete and supervises the job on site thanks to some specialized subcontractors. Due to the current market uncertainties, this solution allow glassmakers to reduce their investment and have a fast track from glass to glass, thus reducing the production stops and losses.

SEFPRO has based this experience mostly on the Insulation Market where several furnaces have been repaired successfully worldwide. Depending on the furnace configurations, the tank walls, the bottom and even the throat can be cast. The lifetime of the furnace after a CIP operation is almost equivalent to what can be achieved with the comparable bonded blocks. SEFPRO is able to repair a complete furnace tank in 3 working days with more than 40 tons cast.

Thanks to the validated experience on the Insulation Market, SEFPRO has developed and is now proposing some new solutions for soda-lime furnaces on Container or Speciality glasses markets. Based on the existing ERSOL concrete, taking advantage of the cement setting, SEFPRO has created a new self-flow product, whose formula contains a major fraction of electrofused AZS grains. This composition exhibits especially good mechanical, and glass contact properties as well as corrosion resistance compared to the existing solutions which are mainly using zircon/mullite components. Unlike the zircon mullite compositions which are rapidly penetrated, transformed or corroded, the AZS cast in place solution is especially resistant. Several lab and industrial tests

207

confirmed this behavior. On the top of that, it can be installed easily thanks to its self-flow behavior. This concrete is dedicated to full or partial repairs of tank walls or superstructure areas. Such a solution has already been chosen by some customers like enamel frits producers which are looking for a fast repair solution and for a refractory material resistant enough to their very aggressive compositions.

SEFPRO is also proposing a high alumina solution for forehearths. Thanks to a dedicated process, a full or a partial feeder can be cast in place. Any shape or feeder radii can be achieved and the process has been especially developed in order to avoid the occurrence of bubbles on the feeder working surface. The length of each single feeder item can be much longer than the classical unit length (610 mm generally). Reducing the number of joints is therefore an advantage regarding glass corrosion resistance and blistering. The corrosion resistance of such a solution is also similar to the standard sintered solutions available.

Those solutions are now available upon request. SEFPRO will be pleased to study any specific demand in order to provide the most suitable solution and is still devoted to delivering the classical electrofused or bonded blocks to get the most efficient package.

CONCLUSIONS

Thanks to a large experience on the Insulation Market with a chromium oxide cast in place solution, Saint-Gobain SEFPRO is now proposing some unique AZS and high alumina concrete solutions for tank walls, superstructure walls and forehearth repairs. Both compositions exhibit particular good mechanical, glass and vapor contact behaviors for the different furnace areas.

Process Control
and Modeling

NON-ISOTHERMAL PENDANT DROPS OF MOLTEN GLASS: PART 1

Byron L. Bemis
Owens Corning Science & Technology
Granville, OH USA

ABSTRACT:

The physics of formation of pendant drops has been extensively studied for many years, however drops formed during the production of glass fibers present a challenging twist to the classic predictions of constant property axisymmetric drop formation. In the case of melt spinning of silicate glasses the process starts from the formation of a pendant drop and its subsequent detachment from the capillary, where under the correct conditions it creates a trailing fiber as it falls. Thermal and flow conditions on the drop vary substantially in the azimuthal, radial and axial directions, which coupled with the temperature dependent properties of glass melts, create a unique departure from simple pendant drop formation and breakaway. These deviations are important to those designing fiberizing processes. Before embarking on a full three dimensional simulation, a study focused on a two dimensional axisymmetric simulation and its comparison to analytical and experimental results was undertaken to illustrate the difference between the formation of a non-isothermal glass drop and a constant property drop.

INTRODUCTION:

Drops of liquids have held researchers interest for many years. As mathematical curiosities for famous early fluid dynamicists, the pendant drop from a capillary provided an interesting and practical challenge. Young (1805) and Laplace (1806) independently developed the theory of surface tension and drop formation while the first analytical solutions to their theory were completed by Gauss in 1830 [1]. Much of the early work on pendant drops involved numerous methods involving the determination of drop volume or shape with experimental techniques and using the available theory to determine surface tension of the liquid-gas interface. These methods are well detailed in the works by Adamson[2], Padday[3], and Reed & Hah[4]. The studies of the early researches developed into the rich and diversified field of interfacial fluid dynamics. The advancement of theory and numerical techniques has steadily increased the ability of researchers to better understand and control interfacial behaviors.

For industrial researchers the behavior of fluids issuing drop wise from a capillary tube is a critical process step that requires a high level of understanding to achieve precise process control. Industrial processes utilize drops formed over a large range of formation rates and drop sizes ranging from the microscopic such as inkjet printers and micro-fluidics to mid-scale medical devices to large scale mixing and spraying. The vast majority of the past studies have been on constant property fluids at low temperatures and viscoelastic fluids at constant temperatures.

When considering the formation of a pendant drop of glass into an uncontrolled atmosphere the formation of the drop is significantly different than the commonly studied cases. The formation of glass drops in this manner is of vital importance to the persons responsible for designing fiber forming processes and apparatus. The following monograph will examine the deviation of glass drop formation from the constant property creeping flow formation of drops of water and similar liquids. The primary objective of the study is the prediction of the maximum drop diameter along with identification of the main parameters controlling that diameter. The maximum drop diameter plays directly into the design of glass fiber forming bushings.

FORMATION OF CONSTANT FLUID PROPERTY DROPS:

The formation of constant property drops has been studied extensively and is reported in many textbooks and journal papers. Several text books are available on the subject [5,6]. Eggers has been a leading researcher in jets and drops over the past decade and has published a number of papers [7,8]. The reader is encouraged to review these resources for a more in-depth look at jets, drops, and sprays of common fluids extending to breakup. A brief overview of one method of predicting the maximum diameter of a static pendant drop is presented here for context and to compare with experimental results obtained with common fluids that are used to validate the experimental procedures and the numerical simulations, thereby providing a foundation for the extension of the concepts to non-isothermal fluids with varying properties.

In the simplest case, a pendant drop is forming very slowly from a vertical capillary tube into an open air environment. Figure 1 depicts the drop being considered. In this simple case, the geometry is axisymmetric, which can be used to great advantage. The familiar Navier-Stokes equations in cylindrical coordinates are the natural choice for describing the problem at hand under the following assumptions. The fluid is isothermal and has constant viscosity (μ), axisymmetric ($all \ \frac{\partial}{\partial \theta} = 0$) and the gravity force (g) acts only in the z-direction. The velocity (U) with components (u_z, u_r, u_θ), the density (ρ), and the capillary radius (r) describe the flow.

Continuity:

$$\frac{\partial u_z}{\partial z} + \frac{1}{r}\frac{\partial(r\,u_r)}{\partial r} = 0 \tag{1}$$

Momentum:

Radial Component-

$$\rho\left(\frac{\partial u_r}{\partial t} + u_r\frac{\partial u_r}{\partial r} + u_z\frac{\partial u_r}{\partial z}\right) = -\frac{\partial p}{\partial r} + \mu\left\{\frac{\partial}{\partial r}\left[\frac{1}{r}\frac{\partial}{\partial r}(r\,u_r)\right] + \frac{\partial^2 u_r}{\partial z^2}\right\} \tag{2}$$

Axial Component-

$$\rho\left(\frac{\partial u_z}{\partial t} + u_r\frac{\partial u_z}{\partial r} + u_z\frac{\partial u_z}{\partial z}\right) = -\frac{\partial p}{\partial z} + \mu\left[\frac{1}{r}\frac{\partial}{\partial r}\left(r\frac{\partial u_z}{\partial r}\right)\right] + \frac{\partial^2 u_z}{\partial z^2} + \rho g_z$$

For the pendant drop in figure 1, the equations above are subjected to the no-slip boundary conditions at solid surfaces and the kinematic condition on the free surfaces. The kinematic condition implies that there is no liquid crossing the boundary into the gas phase, or in other words forms a definite boundary between the phases. For creeping flows into an atmosphere of gas with minimal velocities there will be no interfacial shear stress tangential to the surface and the normal stress inside the fluid is balanced by the surface tension as described by the famous Young-Laplace equation.

$$\Delta p = \left(\frac{1}{r_\theta} + \frac{1}{r_z}\right) \tag{3}$$

The mathematical implementation of the dynamic boundary conditions at the free surface is quite involved and an exact analytical solution to these equations is impossible even for the creeping flow in this simplified case due to the free surface. However, typical simplifications and assumptions can be applied and a simple useful approximate solution to predict the maximum drop diameter can be developed. As this is an applied research paper we will continue to move into the practical application and the reader is referred to chapters 1-4 of Middleman's text [5] for the details of the mathematics.

When thinking of the formation of a glass drop, and the ensuing experimental program several considerations need to be made regarding the flow regime of the isothermal fluid case and the full glass drop problem to ensure appropriate similarity of the experiments in both cases. Several well known dimensionless groupings can be calculated based on fluid properties, jet parameters, and geometry that can guide the formulation of the experimental program and shed some physical insight into the problem.

A creeping flow vertical jet is a scenario common to fiberglass forming from a bushing tip with a traditional E-glass. From knowledge of this case, one can calculate the Reynolds number (Re), Weber number (We), Capillary number (Ca), Ohnesorge number (Oh), and the Bond number (Bo). Each dimensionless grouping provides an insight to the physics, with the first three including the effects of flow velocity and the last two only fluid properties and geometry. Each is classically defined below.

$$Re = \frac{2\rho U R}{\mu} \tag{4}$$

$$We = \frac{\rho U^2 R}{\sigma} \tag{5}$$

$$Ca = \frac{\mu U}{\sigma} = \frac{2We}{Re} \tag{6}$$

$$Oh = \frac{\mu}{\sqrt{\rho \sigma R}} = \frac{\sqrt{We}}{Re} \tag{7}$$

$$Bo = \frac{\rho g R^2}{\sigma} \tag{8}$$

Table 1 shows the values of the dimensionless groupings for the base glass case along with one of the cases with water as constant property fluid along with the problem dimensions and experimental conditions representative of those in the study.

Table 1: Dimensionless parameter values

	Temperature (F)	Diameter (in)	Re	We	Ca	Bo
Water	70	0.054	3.5E-01	6.1E-07	3.5E-06	6.4E-02
Glass	2300	0.054	1.5E-04	5.5E-05	7.5E-01	3.2E-02

The Reynolds number compares the inertia term with the viscous term. Here clearly with such small values, the flow is viscosity dominated and inertia does not play a significant role. Reynolds numbers up to ~ 400 were examined for water. Even at higher flow rates the predictions of drop diameter were accurate up to Re~200. The small Weber number indicates that the surface tension dominates over the inertial terms and from experience the We< ~3 indicates that there will be drop formation rather than a continuous jet and with the We<< 1 drop formation is independent of flow rate. A very small Capillary number shows that the surface tension dominates over both the viscous and inertial terms. The Bond number compares the gravitational term to the surface tension and again this is a surface tension dominated problem as the Bo <<1. Finally the Ohnesorge number, like the Capillary number, relates the viscous forces to the surface tension, and with the small Ohnesorge number again indicating a surface tension driven problem. The values in table 1 show most importantly that the flow rates and capillary tube diameters chosen provide a surface tension driven non-inertial flow in each case.

As the development of the glass drop is extremely slow, one can also consider the drop as quasi-static and perform a force balance on the constant property drop after Middleman[5]. The volume of the drop having sufficient mass necessary to just balance the surface tension will have the maximum diameter. Consider the drop hanging from the inner diameter of a capillary tube as shown in figure 1. The angle θ, formed by the line tangent to the free surface at the capillary tube exit is used to determine the vertical component of the surface tension force.

$$F_\sigma Cos\theta = \sigma\pi D_i Cos\theta \tag{9}$$

The vertical component of the surface tension force must be equal to the gravitational force for the drop to be in equilibrium.

$$\sigma\pi D_i Cos\theta = \rho g V \tag{10}$$

Solving for the volume, and remembering that this volume is the maximum volume that can be suspended by the surface tension force allows the volume of an assumed spherical drop to be substituted in for V_{max}.

$$V_{max} = \frac{\sigma\pi D_i}{\rho g} \tag{11}$$

$$V_{drop} = \frac{\pi D_d^3}{6} \tag{12}$$

Rearranging and solving for the diameter of the drop leaves:

$$D_d = \left[\frac{6\sigma D_i}{\rho g}\right]^{\frac{1}{3}} \tag{13}$$

Substitution of the Bond number as defined above and arranging to provide a ratio of drop diameter to capillary diameter is a useful equation.

$$\frac{D_d}{D_i} = 1.82\, Bo^{-\frac{1}{3}} \tag{14}$$

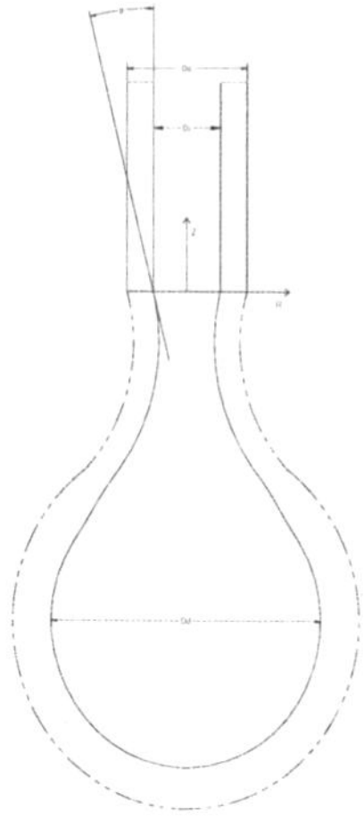

Figure 1: Drop Schematic

It is well known that the assumption of a spherical drop leads to an error that can be empirically corrected to account for the effect of the neck region of the actual non-spherical drop. The correction of 1.82 to 1.6 is recommended.

$$D_d = 1.6\, D_i\, Bo^{-\frac{1}{3}} \tag{15}$$

Applying equation 15 to drops of room temperature water and Advantex[TM] E-glass at 2300 F yields the predictions presented in figure 2.

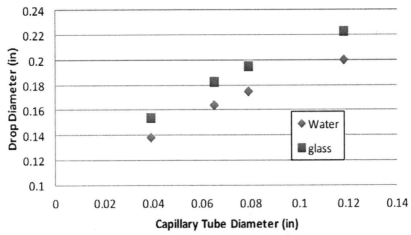

Figure 2: Constant property maximum drop diameter

The constant property predictions are unlikely to be very accurate when the fluid's viscosity exhibits strong temperature dependence and the drop experiences heat transfer from the free surface. Even though the flow rate is very low and inertial effects are negligible, the fact that the drop is forming while properties are changing with time due to cooling leads one to expect that the balance of forces should shift resulting in a drop that should have a different shape than the constant property case. The nonlinearity induced from the variable properties precludes an analytical solution leading us to an experimental and numerical investigation of the formation of the non-isothermal drop.

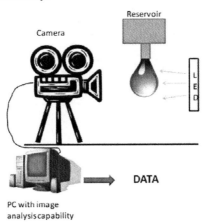

Figure 3: Experimental apparatus

EXPERIMENTAL INVESTIGATION OF DROP FORMATION:

An experimental program utilizing a high resolution video camera to image drops formed from a capillary tube was designed and constructed to test the predictions of the force balance method of predicting maximum drop diameter for both water and glass. A schematic of the rig is shown in figure 3. Drops were produced and imaged, then measured via the image analysis software PCC 2.1 from Vision Research. Image resolution is approximately 10 microns per pixel yielding high accuracy in determining the diameter of the drops from the images.

Figure 4 shows a sequence of images depicting the formation of a water drop through maximum size to separation that is representative of the constant property fluids. Experimental data for water drops taken from various sizes of capillary tubes are presented in figure 5.

Figure 4: Sequence of images of a water drop through separation

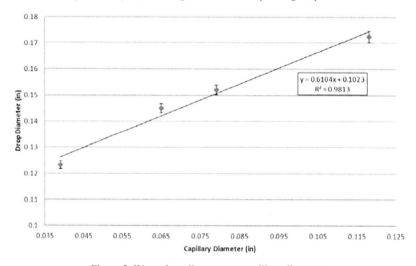

Figure 5: Water drop diameter vs. capillary diameter

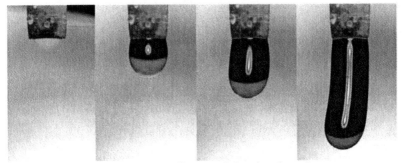

Figure 6: Sequence of images of glass drop formation

Figure 6 depicts a typical non-isothermal drop formed from molten glass. Note that, as expected, the glass drop has a significantly different shape than the water drop shown in figure 4. The drop begins hemi-spherical (shown in the first frame of figure 6) similar to the water, but as the drop Additionally the diameters were formation progresses and cooling drives up the viscosity creating resistance to the tendency towards a spherical shape normally driven by surface tension forces. The drop becomes more of an extrusion. Eventually the glass solidifies and the drop diameter becomes constant. The time to eventual drop separation is quite long.

As only one tube size was available for the glass drop experimental rig, the initial glass temperature was varied to illustrate the effect of the highly temperature dependant viscosity on the diameter of the glass drops. The results from image analysis are shown in figure 7. confirmed by measuring the solidified drops with a micrometer and correcting for thermal expansion. These simple experiments provided the basis for validating numerical simulations which permit a much more detailed investigation of the nature of the non-isothermal drop formation.

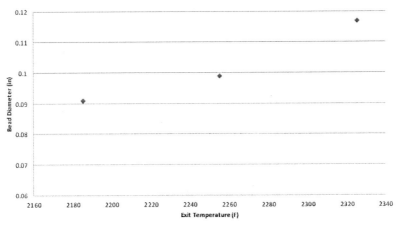

Figure 7: Measured glass drop diameter vs. inlet temperature

NUMERICAL SIMULATION OF PENDANT DROP FORMATION:

Numerical simulation of the formation of drops was also carried out so that the nonlinear effects of temperature dependant viscosity could be further investigated. Water was simulated first in order to test the numerical simulations against the predictions for a static drop and the data from the experiments. The 2D-Axisymmetric simulations were carried out using Ansys Polyflow 14.0. Constant properties were used for water and temperature dependant viscosity used for molten glass. Second order time discretization utilizing the Crank-Nicholson scheme was necessary along with quadratic shape functions for coordinates, velocities, and temperature. Adaptive remeshing was used to maintain mesh quality during the formation of the drop. In the case of the glass drop, the thermal boundary conditions were applied as mixed convection and radiation. Mesh independence was verified, and it also became clear that the T-Grid mesh quality > 0.6 was necessary to achieve accurate simulations. The three data sets are in excellent agreement for water as shown in figure 8. The ratio of drop diameter to tube diameter ranges from 2.65 down to 2.35 for the experiments conducted, decreasing with increasing tube diameter. Figure 9 shows a series of images taken from the simulation of a water drop forming very slowly shaded by velocity. Note the excellent match to the shape of images of the experimental drops.

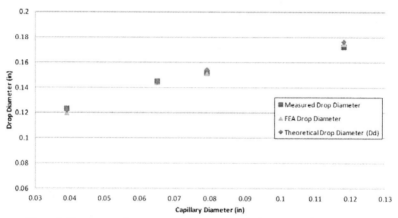

Figure 8: Data comparison, measured data, numerical and analytical predictions

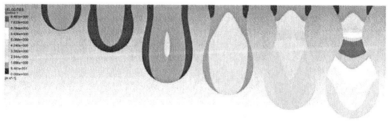

Figure 9: Sequence of images of a water drop forming

Numerical simulations for the glass drop provided a means to expand the investigation beyond what was possible experimentally as well as provide access to all the difficult to measure characteristics of the very hot flow of molten glass. Figure 10 shows a representative sequence of images showing the drop growth with time. Again the shading is by velocity. Notice that the velocity field is significantly different in the case of the glass drop. Where the water drop remains fluid throughout its evolution to separation the glass drop does not. By the time the drop shape begins to develop the classic pendant drop shape (3rd from the left in figure 10) the viscosity has risen sharply.

Figure 10: Sequence of images of a glass drop forming

This large increase in viscosity serves to restrict further movement of the free surface, effectively freezing in the maximum diameter earlier with respect to separation than for the water drop. The maximum diameter is smaller than for the constant property case leading to a much elongated drop shape as commonly seen during restarting Fiberizing bushings. Also note that the velocity

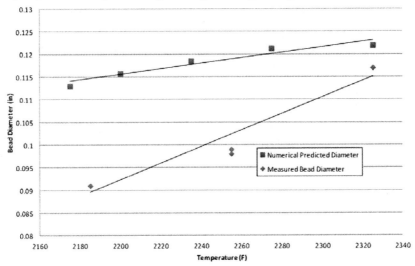

Figure 11: Measured and numerically predicted drop diameter vs. temperature

in the lower part of the drop is constant in the last three frames of figure 10 corresponding to rigid body movement. Figure 11 shows a comparison of numerical experimental results for maximum drop diameter. The deviation in the predictions from the measured values ranges from 20% at the lower temperature down to 7% at the high temperatures.

The numerical results obtained for the prediction of the drop maximum diameter for water were significantly better matched to the experimental values than that obtained for glass. Investigation into the drop shape evolution showed that the simulations provided a very good match to the drop early in the process. Figure 12 shows the drop shape, in white with grayscale velocity vectors, superimposed over the image of the glass drop forming. The numerical simulation

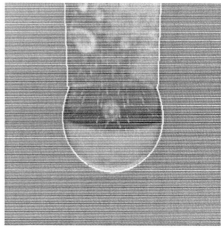

Figure 12: Drop shape predicted vs. experimental at 10 seconds.

reproduces the shape with good fidelity. As time progresses the drop shape prediction begins to deviate from the images. It is clear that there are uncertainties in the correlations used to specify the temperature dependent viscosity and also the surface heat fluxes. In this simulation the effective thermal conductivity of the glass is approximated to include the effects of internal radiative transfer and the weak temperature dependence of surface tension is neglected. These uncertainties and assumptions along with the assumption of constant mass flow rate are the likely contributors to the error and will be addressed in future studies as the limitations of the software are lifted and more precise measurements of temperature dependant properties are under taken. Although there is significant error at the low end of the temperature range, the interest in predicting drop diameters is in the forming range for glass fiber production at about a viscosity of 100 Pa-s which is at the end of the temperature scale where the errors are the smallest.

Further investigation of the effects of temperature, heat transfer rates, mass flow rates, and tube diameter were undertaken using Polyflow. A better understanding of the sensitivities of the problem can be used to guide future design and provide direction to future research in the area. One of the primary interests is of course the effect of tube diameter. In figure 13 the predictions for maximum drop size at a constant flow rate and thermal conditions are presented as a function of tube exit diameter.

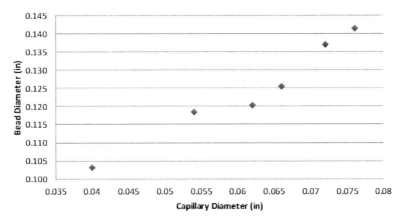

Figure 13: Drop diameter vs. capillary diameter

As one might expect, and as seen with water, the maximum drop diameter increases rather linearly with increasing capillary tube diameter. During these simulations the drop was considered to be pinned to a location at the outside edge of the tube exit as was observed after wet out during the experiments. The dynamic contact line was not simulated in these results.

Figure 14, shows the effect of increasing initial temperature just upstream of the outlet of the capillary tube. The maximum drop diameter increases with increasing temperature. The higher temperature reduces the viscosity in the initial drop and allows the surface tension to

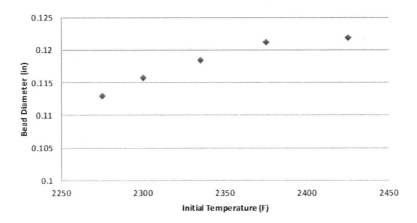

Figure 14: Drop diameter vs. initial temperature

achieve a balance with a more spherical shape before the heat transfer takes over and cools the drop enough to stop the diametrical growth and transition to an extrusion. This leads to a smaller ratio of drop diameter to tube diameter than seen in the constant property case.

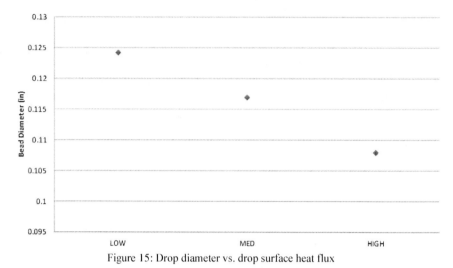

Figure 15: Drop diameter vs. drop surface heat flux

Figure 15 shows the effect of increasing the heat transfer rate from the tip wall and the bead surface. The heat transfer rates were set at three levels characterized by Low-Medium-High.

With low near the lower end of the potential conditions seen in fiber forming and the high values

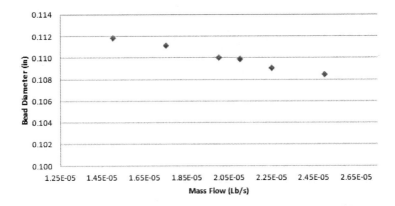

Figure 16: Drop diameter vs. mass flow

corresponding to a fast fiber forming process, but not unrealistically high. Cooling of the tip wall produces the same effect as lowering the inlet temperature, and increasing the heat loss from the drop free surface accelerates the transition from normal spherical drop growth to the more extrusion like growth. The earlier transition leads to an elongated drop of smaller diameter. Here the mass flow was held constant, which in retrospect should have been allowed to vary as the effects of temperature were exerted on the free surface. Future work will be conducted with a more realistic constant pressure inlet condition.

The response of bead diameter to increasing mass flow rate with a constant capillary tube diameter is counterintuitive. Here at a constant temperature the bead diameter decreases with increasing flow rate as shown in figure 16. As flow rate increases, there is insufficient time for the drop to expand diametrically and the bulk movement pushes the drop away from the hotter inlet allowing the surface to cool quicker than in the slower flow cases. Although the effect is clearly discernible, it is a rather weak dependence compared to the changes in initial temperature, heat transfer, and tube diameter.

Water Drops   Glass Beads

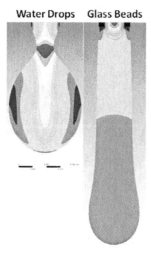

Figure 17: Drop shapes

SUMMARY:

Prediction of the maximum drop diameter for non-isothermal fluids with strongly temperature dependant material properties requires an understanding of the physics of the forming environment, a good characterization of the material's physical properties and an appropriately detailed numerical simulation. Figure 17 illustrates the characteristic drop shapes of the two extremes of drop formation shaded by velocity, which further illustrates the difference in the flows at hand. Where the water is much more spherical in shape and has significant velocities throughout the drop up until separation, the glass has frozen at the leading end of the free surface and thereby exhibits rigid body movement of that part of the drop and the longitudinal growth is through an extrusion-like process. The ratio of maximum drop diameter to tube diameter is larger for the high surface tension constant property cases than it is for the glass drops as clearly shown in figure 17.

Glass drops are clearly a surface tension dominated flow, however with the addition of high cooling rates the rapid increase in viscosity results in arrested diametrical growth of the drop. Longer drops with a much smaller maximum diameter than the isothermal case are formed.

Reasonable diameter predictions are possible using a 2D axisymmetric simulation where the effects of heat transfer and temperature dependant material properties are properly considered. These predictions can be utilized to better understand the interaction of drops or "beads" formed when designing a new forming bushing or developing new processes with new glass formulations or substantially differing forming environments. Although many of the relationships when taken independently are nearly linear in response, when taken all together in the full realistic simulation or the actual process they become interacting and highly non-linear. The solutions are quite sensitive to a number of nonlinear interacting parameters such as heat transfer coefficients, temperature, material properties, and mass flow rate. Further simplification of the problem in terms of simple geometric terms and temperatures is unlikely.

FUTURE RESEARCH:

In many cases the axisymmetric conditions do not hold true, especially when considering heat transfer. Many of the drops "see" azimuthally varying heat losses due to varying boundary conditions leading to a fully 3D situation. We see these effects manifested in curled drops.

Initial 3D simulations with "patched" boundary conditions prove to be extremely slow to converge, requiring very long simulation times, further refinement of methodologies and the solver are likely required to proceed.

A significantly better understanding of the dynamic contact line on the capillary tube land leading to more accurate simulation of the early stages of drop formation should also improve accuracy of drop size predictions. Initial numerical simulations of the dynamic contact line have been interesting and will continue to be refined.

Radiation in the transmitting and emitting media of the molten glass is a significant contributor to the overall heat transfer and cannot be accurately simulated by an effective thermal conductivity at this small size scale. The addition of a full discrete ordinates (DO) treatment of the radiative heat transfer should also improve the accuracy of the free surface shape development, but at a very high cost for simulation speed and resources required.

ACKNOWLEDGEMENTS:

I'd like to thank the GMIC for accepting this work for presentation at Glass Problems and Dr. Purnode and the management of Owens Corning for providing the resources and time to conduct and present this research. Thanks to Fred Grube for all the assistance in the laboratory conducting the experiments and to Hossam Metwally of Ansys for assistance with the initial problem setup in Polyflow.

WORKS CITED:

1: Mobius, D. &. (1997). *Drops and Bubbles in Interfacial Research.* Amsterdam: Elsevier Science.
2: Adamson, A. (1982). *Physical Chemistry of Surfaces.* New York: John Wiley & Sons
3: Hah, C., & Reed, R. (1983). A Method of Estimating Interfacial Tensions and Contact Angles from Sessile and Pendant Drop Shapes. *Journal of Colloid and Interface Science* , 472-484.
4: Padday, J. (1971). The Profiles of Axially Symmetric Menisci. *Phil. Trans. Royal Society of London* , 265-293.
5: Middleman, S. (1995). *Modelling Axisymmetric Flows: Dynamis of Fils, Jets, and Drops.* San Diego: Acedemic Press.
6: Clift, R., Grace, J., & Weber, M. (1978). *Bubbles, Drops,and Particles.* New York: Acedemic Press
7: Eggers, J., & Villermaux, E. (2008). Physics of Liquid Jets. *Reports on Progress in Physics* , 1-79.
8: Eggers, J. (2005). Drop Formation-An Overview. *Mathematical Mechanics* , 400-410

ICG TC21 MODELING OF GLASS MELTING PROCESSES - HOW RELIABLE AND VALIDATED SIMULATION TOOLS CAN HELP TO IMPROVE GLASS MELTING EFFICIENCY AND PRODUCTIVITY.

H.P.H. Muijsenberg
Chairman of ICG TC21
Glass Service Inc, Vsetin, Czech Republic

ABSTRACT

Mathematical modeling of glass furnaces started around 1965. The question is what can these models do and how reliable are the predictions of such models? In 1990 the ICG (International Commission on Glass) started the TC Number 21 focusing on "Modeling of Glass Melting Processes". The aim of TC21 is to share and exchange current practice and to develop the theory and application of mathematical modeling of glass furnaces. The activities of TC21 are often carried out as round robin tests where model results of members are compared to each other and in some cases with the actual measured data. A step wise validation of different components of the models related to the whole glass furnace is undertaken. The idea is to come up with improvements to improve the mathematical modeling of each member. The paper will show some validation experiments carried out by several authors over the years within TC21, but also without. These validations show a fairly good agreement between measurements and models. Certain errors are more likely to come from unknown glass properties and boundary conditions, than from the mathematical model itself. As example we show the error that can be caused when we do not know the glass properties very well.

1. INTRODUCTION

Modeling has become a powerful tool in the past 10-20 years with improved algorithms to predict operational conditions in glass melting furnaces, forehearths, regenerators, and annealing lehrs, etc. Furthermore, improvements in computer processing speeds have enabled the modeler to work from conventional computers rather than the exotic workstations of the past. However one needs to understand what "it can do" and what "it cannot do". Someone must understand these limits, such that the interpreter is not left with a choice of "Believe it or not".

If we look back in history we can see that simple 1D and 2D furnace modeling started around 1965, so modeling has existed already for over 40 years, however only since 3D glass furnace modeling about 20 years ago the modeling started to be used on larger scale.

To date, modeling has been asked to interpret furnace operating conditions, and to predict glass quality parameters. In this regard, models have been used successfully to evaluate alternate furnace designs, as well as to model different operating combinations of combustion processes and electric boosting, or even all electric melters. Additionally, glass quality parameters have also been modeled quite well by evaluations of the glass melting temperatures, and the fining properties of the glass.

Limitations to modeling may include imprecise glass properties at high melting temperatures, or specifically the dynamic furnace processes such as batch charging, or furnace corrosion. Furnaces by their very nature are dynamic processes that are subject to multi-operational influences that cannot always be modeled.

Once you have a close approximation of reality, you can change input conditions whereby the model can predict the revised process trend.

One should not forget that the model is just an approximation of reality, which highly depends on temperature dependent glass properties. If the utilized glass properties are wrong, then the trend of the prediction will be wrong too. When measuring glass properties such as viscosity, density, electrical conductivity, and thermal conductivity - the measurements will probably contain some errors. This is even more sensitive when measuring thermodynamic properties such as gas solubility or diffusivity where the error can be of an order of magnitude.

The International commission on glass (ICG) has started a Technical Committee Number 21 (TC21) focusing on "Modeling Glass melting processes" to improve the reliability of mathematical furnace modeling. The TC21 has often used so called round robin tests that allow the members to compare their modeling tools and from this see what should be improved. At present, TC21 started to work upon the so called Round Robin Test number 5. In this paper we will show some of the results achieved in the Round Robin Test No. 4 and also show the definition and some results of Round Robin Test No. 5.

Furthermore, the paper will show what modeling can do by showing some specific examples. Subsequently, we will use a container furnace modeling example and vary the glass properties and show how such variations will change the prediction.

The interest in mathematical modeling is still growing. Especially in recent years it seems to have reached the status of being a proven technology. The increased interest in mathematical modeling represents recognition that modeling can be an effective tool in meeting challenges posed by the problems of glass melting. Glass producers have to reduce costs, satisfy quality requirements, be flexible and meet environmental constraints. The mathematical model can help to predict the effect of the new furnace design on these factors.

Even more, since about 10 years of utilizing these 3D mathematical furnace models, even more specialized models for automatic model predictive control are being used to control the glass furnaces on line instead of by operator control.

There is another strong driving force for mathematical modeling, which is the increased computation power. About every year the computation speed, expressed in Million Instructions per Second (MIPS) is doubled. This means that in the last 10 years the power of computers increased by about a factor 1,000. This enabled the modelers to model much more complex situations. Today, even normal desktop PC's can be used for calculations. For example, today it is normal to calculate the combustion space together with the glass melt. It was only in 1990 that the first attempts to do this were made. This brings enormous potential for the future, as most 3D furnace models are now parallelized. Nowadays you can get quad core CPU's for the price of one normal CPU just a few years ago. This already easily give you 2 times 4 processors in one simple PC today that speeds up the calculation times close to a factor 8, and therefore makes the applications more powerful.

The big question for everybody is how reliable is such a model? Can we base our decisions for the investment of millions of Euros/Dollars in these furnaces, on the outcome of such a mathematical model? In this paper we will try to help you to make your own conclusions, and if you can believe in these mathematical models, or not!

2. FURNACE MODELS IN GLASS INDUSTRY
The paper will focus upon modeling of the glass melting process and the combustion space from the doghouse until the delivery of the glass to the forming process. In Figure 1, only the glass flow in the melter is illustrated. One important result of modeling is the recirculation or the flow of the glass in the melter. The continuum process models describe these processes in terms of the well known equations of continuum mechanics (e.g. Navier-Stokes, differential temperature and balances of mass and species) and phenomenological laws describing the relationship between flux and gradients (e.g. Newton's law of viscosity, Fourier's law of heat conduction, and Fick's law of diffusion). Of-course the process models must also include mathematical representation of other impacts on flow, heat and mass transfer (e.g. electric heating, bubbling).

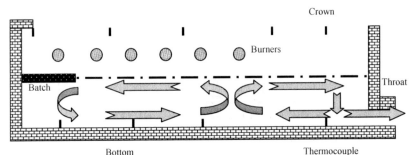

Figure 1. Simplified view of glass flows inside glass melt.

These models are usually divided in primary models, which solve the temperature and flow in the glass and usually a separate model for the combustion atmosphere including combustion and radiation. The secondary model calculates quality, by tracing sand grains, bubbles, stones or in the combustion space e.g. NOx or SOx. Figure 2 gives an example of temperature and glass flows in a flat glass melter.

The history of modeling glass furnaces covers around 35 years. The first attempts of modeling the melting phenomena by mathematics in glass furnaces started around 1965 by Mr. W. Trier [1] from the HVG institute in Frankfurt Germany. See also other references for more information [1-12]

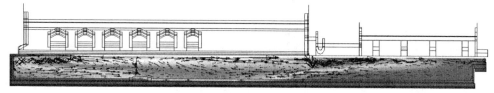

Figure 2. Glass temperature and glass melt flow example in a flat glass melter. Note the recirculation of the glass melt coming back from working end into the melter (by Glass Service).

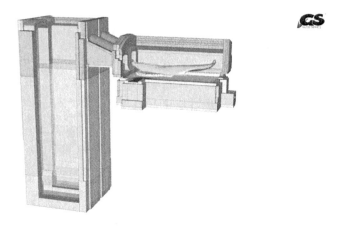

Figure 3. Example of a complete modern furnace simulation model including regenerators (by Glass Service)

3. VALUE OF MATHEMATICAL MODELS

So, as you can see mathematical modeling of glass furnaces has a long history and there are several companies or groups who can help you, or offer you such a service. You probably still have the question, can you trust results of these models and which model should I choose? Which one is reliable? What is the difference?

Basically, there is or should be almost no difference(s) from these different models! Afterall, all models are based upon the same principles of solving the same equations of mass and energy, and thus we may assume that if one is able to program the model without mistakes, the outcomes should be the same. But this assumption might be a problem. As a case in point, you can see that Microsoft with their professional programmers is even not able to program without mistakes. So probably, and unfortunately, all these technical glass model programmers will probably make mistakes too. Only we do not know if they will make a critical mistake or just a minor one. So we need validation of the model. That means that we need a comparison of the calculated results with known or measured values, and it can be very helpful to check if there are not some serious mistakes. Another problem arises, at temperatures of 1500 °C, as we cannot measure so much the glass properties, and what we can measure is not always reliable. We will discuss this in more detail in Chapter 4.

The other question becomes, what is the real value of the outcome of these glass furnace models? This is not only, can I trust that temperature X at spot Y is Z degrees? It means how fast, reproducible and easy can I receive the results?

Let us take the example of a carpenter. The carpenter uses a tool: The hammer. We assume that he has validated and checked this hammer before. But how your house will look like, does not only depend on the quality of the hammer. It depends more on the skills of the carpenter. Does he know the behavior of the wood? Can he listen to you about what the house should look like? Can he read your drawings? Can he make you some recommendations based upon his experience before starting? How fast can he do it? How nice will the house look after it is finished? The hammer is important, but the final results depends more upon the carpenter's knowledge, experience and skills.

This is also valid for somebody modeling your glass furnace. The mathematical model has to be reliable, but the real outcome depends today more on his knowledge of glass properties and glass furnaces in addition to the mathematics. To start some modeling, he will have to use some measured glass properties, or if not available take them from his database or even make some estimates. Not all boundary conditions are known exactly. Sometimes he will have to estimate some heat-loss, because the walls are corroded for instance.

So the value of the outcome of the mathematical modeling study depends on:

1. The code (accuracy, coupling error free)
2. Speed of the code (answers this week not after half a year)
3. User friendliness (how fast can one setup a case, also reducing errors)
4. **Glass properties** (measured, database or estimated)
5. Boundary conditions (measured or estimated)
6. Experience of modeling engineer (does he know that a glass furnace is hot?)
7. Interpretation of the results (not just colorful pictures)
8. Impact upon the glass melting performance (e.g. quality, bubbles, stones etc.)
9. Post-processing (can you understand the pictures, and recognize your furnace?)
10. Costs and time.

As you can see the code itself just 1 out of 10 points. But yes, the carpenter is also not able to build your house without the hammer. In this paper we will demonstrate what the effect can be on the relative prediction when we do not now the properties very well.

4. VALIDATION OF MATHEMATICAL MODELS

So let us see how one can validate the mathematical models, and if it is possible? The models come up with temperatures in the glass melt, refractory, combustion space and for instance exit temperatures of the glass and waste gases. Next to this, the glass models calculate the speed (flow) of glass in the melter and working end. Last but not least, the real glass furnace produces a certain glass quality, and hence some level of defects, and for instance bubble (seed) defects. These bubbles contain certain gases inside which can be checked too.

TEMPERATURES (ERROR DISCUSSION)
Let us first analyze the temperatures. Temperatures can be measured directly by thermocouples and indirectly by optical pyrometers. The direct temperature measurement with the thermocouple can be "in the glass" or "in the refractory". Depending upon which type of thermocouple one uses, the absolute temperature error can be up to about 10 °C. A typical vertical gradient in the glass melt is about 3 °C/cm, which means that the position of the thermocouple is important. It is easy to make a temperature prediction error for the thermocouple position by a few centimeters. The other problem is that the direct measuring thermocouples usually do not last very long and give wrong values, whereby the derivation could be as high as 50 °C, and in some cases 100 °C.

When the thermocouple is inside the refractory its life-time reliability is longer, but the absolute temperature is less reliable. A typical temperature gradient in the refractory can be 10 to 20 °C/cm. Thus an error of more than 20 °C is not uncommon.

In the combustion space, thermocouples usually receive radiation from all around and do not give a good representative value of the gas around them. Typically a good example of this error we can see is for instance just above the regenerator checkers where a thermocouple is exchanging radiation heat with the top of the blocks and usually gives a wrong reading for the preheated air or waste gas. So in the combustion space, one should use in fact, suction pyrometers.

Also, even optical pyrometers have some problems too. When they measure refractory temperatures, one has to set and estimate the emissivity of the refractory material. Maybe the user knows the emissivity as measured in the laboratory, but during operation a glassy layer that has settled down on the refractory and can influence it. Usually you see that 2 different operators will achieve 2 different values. Measuring the glass (surface) is even more critical and it depends highly upon the utilized wavelength of the pyrometer. Depending upon the wavelength one measures, a certain (small) depth into the glass can occur.

GLASS FLOW
Secondly, we still have the speed of the glass flow. This is even more difficult than temperature, but still some techniques are available.

FLOATERS
The first technique is measuring glass surface flows with the aid of floaters. These floaters are usually submerged a little into the glass melt and they should follow the speed and direction

of the glass surface flow. Of course, there are some problems. The first problem is to have good access into the furnace to introduce the floaters. Sometimes people also have made holes into the crown to throw them from the top, for instance. Second, we need (good) visual access from several angles into the furnace at the same time to follow and register the position of the floater inside the furnace as a function of time. This is very difficult, as usually from each peephole the vision angle is limited and the dimensions are distorted by the lens effect (density difference) of a peephole. One can easily imagine errors of 10-50% for the speed. The third complicated factor is the effect of the flames or combustion gases on the floater and glass surface. If you have ever looked to a video of the batch movement at high speed, it should be clear to you that in most cases the flames have indeed a large effect (by friction) on the local flow of the glass surface. In an end fired furnace this really can result in a floater moving in the opposite direction (pushed by the flames) than the glass flow. This can be easily be checked by the effect of the floater after reversing the firing side of the furnace. Even here, big errors can be expected here.

TRACING

The second technique used, is the tracing of special components that can be measured/detected in the glass furnace after melting. In the past, sometimes some radioactive tracers have been used, but today, this is usually not allowed because of safety reasons. Hence, another popular method is for the addition of some tracer material, such as zinc oxide (ZnO). This usually has no or a minor effect on the glass properties and can be easily detected in the glass samples later. This ZnO is introduced during batch mixing or directly into the doghouse. Then the ZnO will be melted with the batch and follows the glass currents. Some ZnO will show the fastest or shortest residence time in the furnace and sometimes one even can recognize several peaks identifying several recirculation loops in the furnace. A problem here is mainly in the way of introduction or mixing the tracer material. This can lead to an error of about 0.5 or 1 hour. The other limiting factor is the amount of samples one is willing to analyze, e.g. each 15 minutes during 24 hours. So the resolution time is at maximum 15 minutes, but if the minimum residence time of the furnace is 4 hours then the error can be relative large. The other effect is how long will the glass sample take to come from the end of throat or canal or feeder through the forming process and into the annealing lehr, because one has to correct for this residence time. This tracing can be easily reproduced, by tracing mass-less particles in mathematical models. Comparisons are sometimes within half to one hour [21].

CORROSION PROFILES

Another method is to check the corrosion profiles on the bottom or side walls of the furnace. This can confirm flow directions that occur close to the refractory. This was presented in one paper from Chmelar and Schill in 1993 [13]. They not only showed a good agreement with floaters on the glass surface but also between corrosion profiles on the bottom and calculated glass flows.

MEASUREMENTS AND COMPARISON

Only a few papers have been published showing extensive measurements and comparison of the modeling results. In the past, companies like Corning and Philips had high interest in getting more insight inside their furnaces. Both of them used a special technique to measure glass depth temperature profiles in the melters by introducing thermocouples. One introduced thermocouples through the original bottom openings and the other one by introducing them through the original openings in the crown. Here, we will give some example of results that Philips achieved in cooperation with TNO in one of their TV screens melting furnaces. For more details see the paper of Muijsenberg and Roosmalen [14].

Figure 4 shows an example of the comparison between two measuring points in the beginning of the furnace. K1 is still under the batch and K2 is just after batch melt out. The bottom of the furnace starts at 0.3 meter and the glass surface is at 1.3 meter:

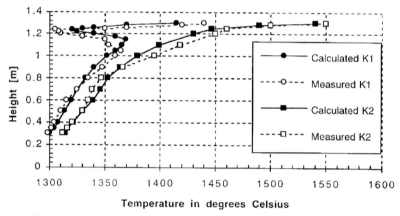

Figure 4. Comparison between calculated and measured glass temperature profile [14]. (by TNO)

So one can see here that locally there can be errors of up to about 30 °C, but the general trend shows a very good agreement between the measurements and calculations.

Chmelar, et al [15] shows the contribution from 1997 of a comparison between the calculated and measured breast wall temperatures of an oxy-fuel furnace. In this case the temperatures were measured by an optical pyrometer. These results show the agreement between the temperatures in the combustion space as well as the effect of the glass surface temperatures on this breast wall. Besides one peak, which is not shown in the model, the model shows the biggest error near the batch blanket area, which is up to 40 °C. See Figure 5.

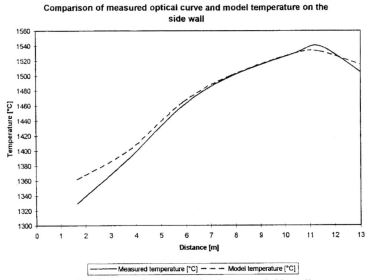

Figure 5. Comparison between the calculated and measured side wall temperatures [15].

5. TECHNICAL COMMITTEE 21 "MODELLING OF GLASS MELTING PROCESSES"

Measurements of the glass velocities and temperatures in glass furnaces are not easy and are limited. Therefore, an alternative method for checking and comparing models was started. Within the International Commission on Glass (ICG) it was decided to start a special Technical committee 21"Modelling of glass melting processes" to explore the reliability of glass furnace models. This committee was for some time Chaired by Mr. Muschick (Schott glass) and Mr. Muijsenberg (Glass Service) as Vice Chairman. From about the year 2000 until 2008 it was chaired by Mrs. Onsel (Sisecam). Since 2009, TC21 is chaired by the author of this paper. Within this committee work, it was decided first to model some special defined test case that would be modeled by all participants as a round robin comparison. The results of this first test are described in a publication from 1998 [16].

Figure 6 shows the simple description of this theoretical furnace (actually just a simple box):

Geometry of the simplified glass furnace, including melter and 'combustion space'.

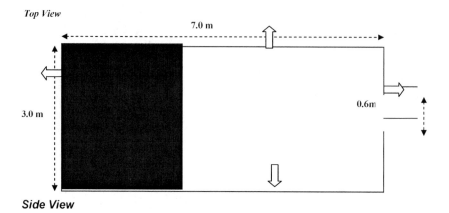

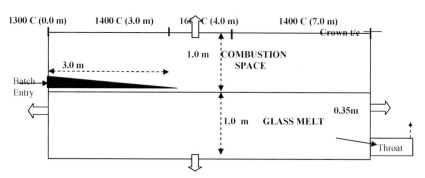

Figure 6: Schematic of theoretical furnace

Figure 7 below, shows some agreement between different companies using different software to model the same defined case.

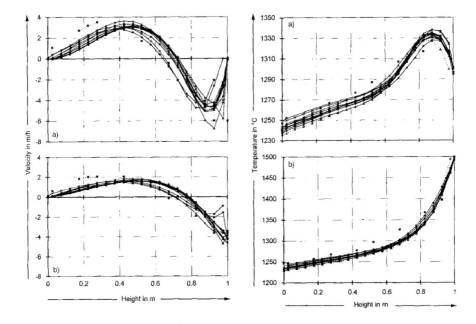

Figure 7. Showing velocity and temperature profiles agreement between different models
for defined test case [16]

From the above figure, you can conclude that the general trend in the glass temperature and
speed is very similar. But if you look to the absolute results upon a certain selected point, you
can still find differences between the models of about 10-15 °C. This is probably due to the
different approach about how different users define and setup the boundary conditions. In
some cases it also might also be due to incomplete or even a not good convergence, maybe in
a few of the models. If we look for instance to the throat temperature we found a spread of
about maximum 15 °C.

Later it was decided (within TC21) to also model an existing furnace with measured data. For
this example, Visteon Glass (Ford) supplied data of their (no longer existing) furnace in
Nashville TN. This furnace was a regenerative fired float furnace in which Ford together with
Corning measured some of the glass temperature profiles. In example, we show you in Figure
8 the results achieved by Glass Service with their own Glass Furnace Model.

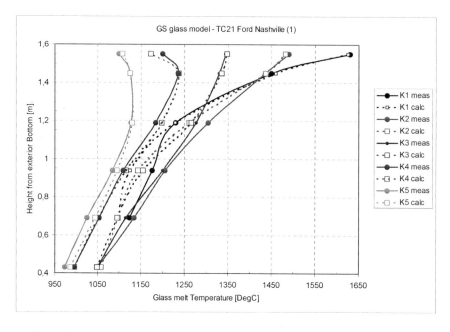

Figure 8. Comparison between measured and calculated temperatures in the former Ford Nashville, TN, float glass melter, calculated by Glass Service.

This figure shows a fairly good agreement especially in the trend. Near the bottom and surface, the temperatures are within 10 °C. Note that these results were achieved by a full coupling with the combustion chamber. In the middle of the glass, however, there can be differences up to about 40 °C, especially in the batch area.

The canal temperature calculated by several participants was within a range from 1,110 °C till 1150 °C. Note that these results were calculated blind, without knowledge of the participants about the results. This shows that in most cases some interaction and fine-tuning of (unknown) input data can be necessary. As in this example, for instance, exact temperatures of preheated air or insulation values of refractory were not known.

Also in Japan such a Round Robin test between several models and one physical model has been carried out. This resulted in good agreements between most participants and the physical model. Chapter 6 will give also such an example.

At present TC21 started to work on Round Robin number 5. The data were gathered by LG Philips Display in cooperation with the TNO Glass Group. The furnace is a former furnace of LG Philips Display that produced TV screens with relative high glass quality.

The definition is summarized below. There are 2 situations to be modeled. One data set is related to March 2004 (Case 1) when the furnace was pulling 236 tons per day and one data set is from April 2002 (Case 2) when the furnace was pulling 220 tpd. During March 2004 the quality of glass was better resulting in fewer bubbles per ton of glass than during April 2002.

The full description can be received from TC21.

Figures 9 and 10 show the temperature distribution in a 3D overview showing the flames as iso-surfaces and a vertical slice showing temperatures as measured and calculated along the height of the melter by Glass Service.

Base Case

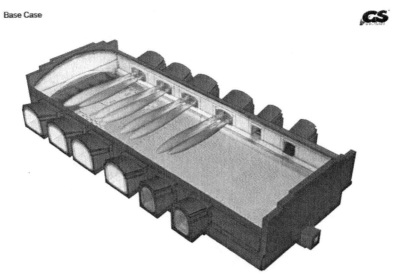

Figure 9. 3D overview of combustion chamber of the TV screen producing furnace subject of Round Robin Test no. 5.

RRT5 - BaseCase - March 2004
Streamlines represent 100 s of glass flow in melter and 0.05 s of flue gas flow in combustion space: Side View (X2)

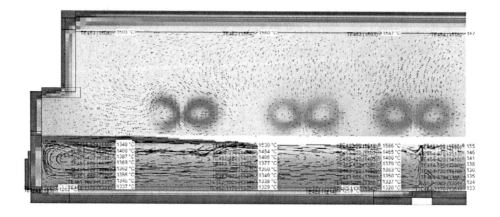

Figure 10. Results of measured and calculated temperatures at selected vertical profiles in the center of the melter from the glass surface until the bottom. On the left side you can see cold batch on the glass surface and above clearly the hot flames radiating into the glass.

In Figures 11 and 12 we can see comparison of measured and calculated temperatures from the different furnace modeling software for the 2004 data.

The blue diamond with the error band is showing the measured depth temperature profile every 15 centimeters.
The blue solid line is the model results of Celsian/TNO with their GTM X model
The Red line with squares is the model results of Glass Service with their GFM 4.12 version
The Green triangles are the model results of Mr. Muschick (Ex Schott)
The Green Crosses are the model results of Sisecam

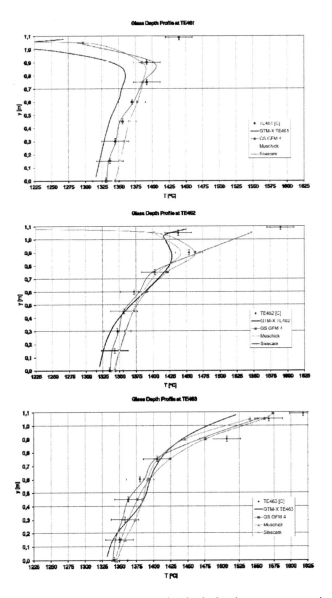

Figure 11. Comparison between measured and calculated temperatures as simple XY graphs
On Y Axis is height of melter on X Axis is temperature in degrees Celsius

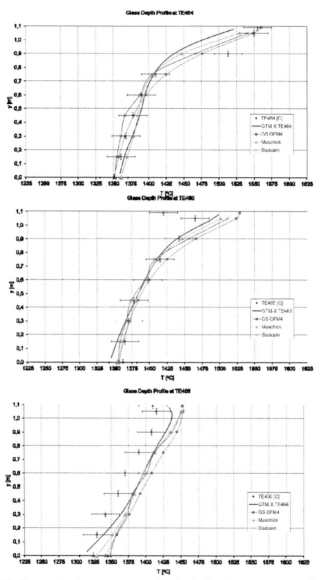

Figure 12. Comparison between measured and calculated temperatures as simple XY graphs
On Y Axis is height of melter on X Axis is temperature in degrees Celsius

We can see a good agreement in most parts of the melter. Under the direct batch area, the temperatures seem to more or less match, however the actual measurements also vary in reality more. Downstream from temperatures 465 and 466, we see the model results are typically warmer. This seems not to be a model error, but actually an error for the definition of the estimated cooling air in Ports 5 and 6. We have to conclude that there has to be more cooling air than it was specified in the input conditions.

In most areas the agreement between the measured data and modeled data for each of the models is within ±15 °C

6. GLASS QUALITY

What kind of melting efficiency can I expect from this new furnace? How many bubbles per kilogram will it produce? Or how fast can I make a product change? These are the questions a glass producer would like to have answered by the modeling. Mathematical models are a very good tool to help to select the best option, but they are not able (yet) to say exactly how many bubbles per kilogram of glass you will get. This is not only limited by the accuracy of the models, but also due to the fact that we cannot know now, how many bubbles per square meter per time unit will be nucleated. This depends strongly on the applied refractory material and also on how the furnace was constructed and heated up. If, for instance, the first bottom lining is damaged during heat up and the patch comes into contact with the glass, this can lead to extensive bubble nucleation. But when the initial modeling was done nobody could expect or foresee this. The good news is that if we assume a certain bubble source and a certain amount of nucleated bubbles, then **Yes** the model can help us to select the best furnace. That means the furnace, which is able to remove most of the bubbles out of the glass before they end up in the product.

This can be done by first calculating the temperature and velocity in the glass melt. Then the redox and gas distribution dissolved in the melt can be determined. As the last item, we need to start bubbles from a particular origin and trace them. During the path travelling, the furnace gases can diffuse into and out of the bubble. For instance, oxygen and SO_2 can rapidly diffuse into the bubble, making the bubble grow and ascend faster to the glass surface and leave the glass melt. With the following example, executed by Glass Service [18] we want to show you how accurate this kind of prediction can be.

Figure 13 shows us the change of an initial Argon bubble as function of time in a soda lime glass. CO_2 (fast) and N_2 (slow) will diffuse into the bubble. Consequently the percentages of gases inside of the bubble and the size of the bubble will change over time. We can see that the bubble reaches a new equilibrium after 3,000 seconds or 50 minutes at 1,500 °C. The figure shows the good agreement of the mathematical model and comparison with an experiment and analyses of bubble composition.

In the furnace model we can trace multiple bubbles with varying sources, location and sizes to see if they end up in the product. If they end up in the product, we can compare their size and composition with measured compositions from bubbles in the real product. Based on this comparison we can conclude what the possible source of the bubble problem is and try to remove or reduce it.

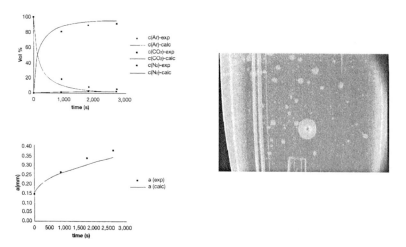

Figure 13. Calculated and measured composition of bubble as function of time [18]

From the above figure and other experiences we conclude that it is possible to predict bubble compositions and so help us to locate the potential source(s).

7. GLASS PROPERTIES

The biggest difficulty for the glass modeler is given by the glass properties. Most glass producers have their own specific knowledge of the glass composition they are melting, but if you ask for glass properties it is usually limited to density, heat capacity and viscosity. One of the most important data for glass modeling is the thermal conductivity, or in fact radiative properties from e.g., Iron, Chrome, and the redox state. This data is usually not known by the majority of glass producers, because it is not vital information they need for the melting and production of glass.

But today it is possible to measure this property at several institutes like TNO, Glass Service, and the Universities in Prague or St. Petersburg [19]. Of course, all data should be supplied as function of temperature. The thermal conductivity is usually presented in a polynomial form to the third order as function of temperature. The error for this can easily be 10%. Next to it, the thermal conductivity in fact varies as function of the local redox state in the glass melt due to the shift of Iron from the 2^+ state to the 3^+ state. Fe^{2+} absorbs heat and Fe^{3+} almost does not.

The modeler needs extensive data for all utilized properties of the refractories and insulations as a function of temperature. That means he needs, for instance, also the density, heat capacity, emissivity and for sure the thermal conductivity as function of temperature. For most refractories this is measured and supplied by the manufacturer. But how sure are we that the refractory that is used in a furnace indeed behaves exactly than the one which was measured once some time ago in the laboratory? And what is the effect of corrosion and penetration of some glass into the refractory pores and in-between stones?

If we want to calculate glass quality we also need properties of certain gases in the glass, like e.g., O_2, CO_2, Ar, SO_2, SO_3, N_2, and H_2O. For these gases we need solubility, diffusivity and the redox equilibria as an (exponential) function of temperature. Next to it, these properties change as a function of the glass composition. The availability and reliability of all these glass properties data is still limited, and is maybe the most important data or part to complete a modeling job successfully [20].

In the following example we will test the effect of a certain error in glass properties upon the predicted temperatures. Figure 14 shows the demo furnace we have been using for a test of the effect of the glass properties. We have increased and decreased the glass viscosity and thermal conductivity by a factor of plus and minus 10%. Then we calculated the effect upon the glass bottom temperatures. With the increased thermal conductivity, the temperatures increased by 6 °C and with the decreased thermal conductivity the temperature decreased by 6 °C.

We wanted to also test what is the change of the trend prediction when we are using the wrong glass properties on a change of 5% fuel increase and bubbling versus no bubbling. A fuel increase of 5% resulted in the base case into a temperature increase of 51 °C. When we calculated this with the wrong glass properties for viscosity or thermal conductivity the relative prediction was just 50 or 52 °C. So based on these results we may conclude that even if the glass properties are not perfect that the simulation models relative prediction is still reliable.

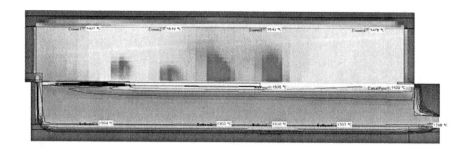

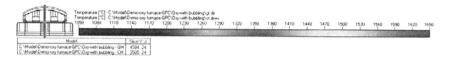

Figure 14. Shows the temperature in the side view of the center plane of a Demo container furnace simulation. The bottom temperatures are in the range of 1300 °C.

8. CONCLUSION

The paper tried to show you the state of the art of glass furnace modeling and its validation. Several mathematical models are available today and probably most of them are validated and accurate. The paper shows that experience, glass properties and boundary conditions have a large influence on the outcome and value of the modeling results. Assuming that thermocouples can show errors, it is fair to expect some agreement between measured and calculated glass temperatures in the range of 10-20 °C. If the difference is larger (at 1 point) then it is more likely that a certain thermocouple is not representative any more, than that the model is wrong. The other explanation of a mismatch can be due to estimated or unknown glass properties. "Garbage in" results in "garbage out". Glass furnace models can be a helpful tool to design a better furnace and to optimize the melting performance of the furnace. It is estimated that today about 1,000 furnaces have been optimised/designed by models.

ICG TC21 has given a great contribution to the development and reliability of mathematical models and probably will continue to do this in the coming 10 years even more.

REFERENCES

1. Trier, W. und Stein, A.: Vereinfachtes Rechenmodell des Warmeaustauches durch Strahlung zwischen Flamme und bestimmten Punkten auf der Glasoberflache in einem Glasschmelzwannenofen, Glastechnische Berichte, 38 Jahrgang,, 1965, page 353-361.

2. Hans-Jörg Voss: Mathematisches Modell zur Abschätzung des Energiehaushaltes von Glasschmelzwannenöfen, Glastechnische Berichte, 48 Jahrgang, Heft 9, 1975, page 190-206.

3. Leyens, G.: Beitrag zur Berechnung zweidimensionaler Konvektionsstromungen in kontinuierlich betriebenen Glasschmelzwannen. T. 1 u. 2. Glastechnische Berichte 47, 1974, Nr. 11, S-251-259; Nr. 12 S 261-270.

4. Trier, W.: Temperaturmessungen im Glasbad von Wannenofen. Glastechnische Berichte 26, 1953, S.5-12.

5. Suhas V. Patanker: Numerical Heat Transfer and Fluid Flow, 1980, ISBN 0-07-048740-5

6. L. Nemec, The refining of glassmelts, Glastechnische Berichte, 15, 1974, page 153-159.

7. A. Ungan and R. Viskanta, Three dimensional Numerical modelling of circulation and heat transfer in a glassmelting tank Part 1. Mathematical formulation, Glastechnische Berichte, 60, 1987, page 71-78

8. Simonis F.: Estimation of redox distribution in the melt by numerical modeling. Glastechnische Berichte. 63K, 1990, page 29-38

9. Schill, P.: Calculation of 3-dim glassmelt flow in large furnaces via twogrid method. Glastechnische Berichte. 63K, 1990, page 39-47

10. Schill, P., Chmelar, J.: Bubbles behaviour in the glass melting tank. In: 2nd International Conference "Advances in the Fusion and Processing of Glass", Duesseldorf, 1990.

11. Carvalho, M. da G.; Nogueira, M.: Physically-based modelling of an industrial glass-melting end-port furnace. Glastechnische Berichte 68C2, 1995, page 73-80

12. Muysenberg, H.P.H.: Modeling the combustion chamber of a glass furnace. Presented at 1st Mathematical Seminar on Mathematical Simulation in Glass Melting, Horni Becva (Czech Republic), 1991.

13. Chmelar, J., Schill, P.; Verification of 3D mathematical simulation of glass melting tank, International congres on glass, Madrid,

14. Muysenberg, H.P.H.; Simonis, F.; Roosmalen, R.; Verification of 3D mathematical simulation with measured temperature profiles during furnace operation. Glastechnische Bericht 68C2, 1995, page 55-62

15. Chmelar, J.; Novackova, M.; Safarik, I.; Budik, P.; Mathematical modeling of furnace design. IV International seminar on mathematical simulation in glass melting, 1997, Horni Becva, Czech Republic, page 140-153.

16. Muschick, W.; Muysenberg, E.; Round robin for glass tank models, Report of the International Commission on Glass (ICG), Technical committee 21"Modelling of glass melts", Glastechnische Berichte. Glass Sci. Technol. 71no6, 1998, page 153-156

17. Nagao, H.; Wada, M.; Three dimensional numerical simulation for visualization of fluid flow, 1990 ?.

18. Ulrich, J.; Nemec, L.; Matyas, J.; Mathematical modelling for the identification of defect bubble sources, International glass review, 1998.

19. Endrys J.; Measurements of radiative and effective thermal conductivity of Glass Service, IV. International seminar on mathematical simulation in glass melting, 1997, Horni Becva, Czech Republic.

20. Klouzek, J.; Determination of the equilibrium partial pressures of sulphur dioxide and oxygen in float glass melt, III. International seminar on mathematical simulation in glass melting, 1995, Horni Becva, Czech Republic.

21. Hermans J.M.; Roosmalen M.P.W.; A tracer trial on a TV-Funnel tank, Proceedings, ICG Annual meeting 2000, Amsterdam.

22. Choudary, M.; A Three-Dimensional Mathematical Model for Flow and Heat Transfer in Electrical Glass Furnaces, IEEE Transactions on Industry applications, Vol 1A-22, No5, September/October 1986.
23. Choudary, M.; Norman, T.; Mathematical modeling in the glass industry: An overview of status and needs, Glastechnische Berichte, Glass Sci. Technol. 70 no. 12, 1997, page 363-370
24. Murmane, R.A.; Johnson, W.W.; Moreland, N.J.; The analysis of glass melting processes using three-dimensional finite elements, International journal for numerical methods in fluids, Vol 8, 1491-1511, 1988.
25. Mase, H.; Oda, K.; Mathematical model of glass tank furnace with batch melting process, Journal of Non-Crystalline Solids 38&39, 1980, 807-812.
26. Hayes, R.; Wang, J.; Mcquay, M.; Webb, B.; Huber, A.; Predicted and measured glass surface temperatures in an industrial regeneratively gas-fired flat glass furnace.

PROPER MODELING OF RADIATIVE HEAT TRANSFER IN CLEAR GLASS MELTS

A.M. Lankhorst[1], L. Thielen[1], P.J.P.M. Simons, A.F.J.A. Habraken[1]
[1]CelSian Glass & Solar B.V.
Eindhoven, The Netherlands

ABSTRACT

In glass melting furnaces and forehearth canals heat transfer is dominated by thermal radiation. To apply Computational Fluid Dynamics (CFD) modeling for enhancement and optimization of furnace and forehearth designs and processes, sufficiently accurate modeling of radiation is therefore required. Two radiation models are generally available in the CFD tools presently applied in industry:

- Rosseland Approximation,
- Discrete Ordinate Method (DOM)

In the Rosseland Approximation, the radiative heat transfer is modeled by a modification of the thermal conductivity, whereas the DOM solves the radiation equation for a finite number of directions. Usually, the Rosseland Approximation is applied in the glass bath and the DOM is used to model the radiation in the combustion space. For the Rosseland simplification to hold, the glass must be optically thick, which certainly is not true for (ultra-) clear glasses and shallow forehearths. Furthermore, the application of DOM for cases with a (locally) high absorption coefficient (e.g. for an optically thick medium) may experience bad (or even non-) convergence and may have results that depend highly on the grid size, especially when the product of absorption coefficient and grid cell size is relatively high.

A significant step forward for radiative heat transfer modeling has been accomplished for GTM-X, CelSian's state-of-the-art multi-physics CFD modeling tool for optimizing glass melting, combustion and forehearth processes. The tool's DOM has been extended to account for spectral variation of the absorption coefficient of both the combustion gases, and the glass melt. Furthermore, it now also features a hybrid option, mainly of interest for the glass bath, making sure that when the product of a (spectral band) absorption coefficient and grid size exceeds a certain threshold, the effect of the (optically thick) absorption is transferred to a Rosseland thermal conductivity.

This paper discusses this hybrid Rosseland-spectral DOM model, applied to an oxy-fuel float furnace. Results are compared to Rosseland and gray DOM calculations, for flint and ultra-clear glass. It is shown that those features of the melt flow that strongly determine glass quality, like hot-spot temperature and residence time distribution, can significantly depend on the applied model for thermal radiation in the melt.

1. INTRODUCTION

Heat transfer from combustion gases and soot to the glass melt as well as internal heat transfer in the glass melt mainly takes place by thermal radiation, especially in the spectral region from $1 - 4$ μm. In modeling studies, radiative heat transfer in the melt is usually approximated by the Rosseland Approximation[1]. This is very convenient, since it allows the radiative heat transfer to be included in the existing transport equation for either the temperature or the enthalpy. Radiation transfers energy between volumes and surfaces that, being close together or far apart, are in each other's line of sight, through the emission and absorption of electromagnetic waves (photons). In the Rosseland Approximation, this non-local phenomenon is approximated by an artificial increase of the local, diffusive heat transport. Thereto, the thermal conductivity of the melt is increased with a 'radiative thermal conductivity' $\kappa_r(T)$, defined by

$$\kappa_r(T) = \int \frac{4}{3} n^2 [1/K(\lambda)] [dM(\lambda)/dT] \, d\lambda \qquad (1)$$

with n the refractive index of the glass melt, M the Planck's function of a black body radiator, and T the absolute temperature (K), and integration is performed over the spectral region in which glass is semitransparent. This Rosseland approximation is valid provided that the mean free path of the photons is small compared to a relevant dimension of the glass furnace. In most parts of the furnace this assumption is satisfied. But it is known that for very clear glasses and for shallow furnace parts, such as feeders, this condition can be violated. In addition, in regions where large thermal gradients exist, such as near the glass surface, the characteristic length scale is not any tank dimension but the thickness of the thermal boundary layer. Therefore, true radiation models, such as the DOM, are believed to yield more accurate results in these cases.

In the DOM, the general radiative transfer equation in a medium is solved by dividing the solid sphere for integration into solid angles, also called ordinates. For each ordinate, a scalar convection equation describes the evolution of the radiative intensity along all parallel lines of that ordinate, from 'upwind' to 'downwind', taking into account any emission or absorption of radiation through surface- or volumetric sources and sinks. This way, the non-local radiative heat exchange is cast into a standardized equation of the convection-diffusion type, which is the basic building block of CFD programs[2]. In gray DOM simulations of glass melts, a constant (wavelength averaged) absorption coefficient κ is used. In spectral DOM simulations the spectral absorption coefficient $\kappa(\lambda)$ is used, obtained from the spectral measurements at high temperatures. Since industrial glasses show absorption coefficients that exhibit spectral variations, depending on the glass composition or redox state, the step from gray- to spectral radiation modeling in the glass melt is expected to further improve temperature predictions.

In GTM-X, the Glass Tank Model of CelSian Glass & Solar (former TNO Glass Group), a spectral radiation model has been developed, both for the glass melt and for the combustion space[3]. In order to accurately model and optimize the heat transfer and melting processes in industrial glass tanks, detailed knowledge of the spectral optical properties of glass melts is required. Presently, experimental methods to determine the high temperature spectral properties of industrially produced glasses are available[4]. If such measurements are not within reach, one can alternatively derive an approximate spectral absorption curve from the glass composition (concentrations of multivalent ions Fe^{3+}, Fe^{2+}, Co^{2+}, Ni^{2+}, Cr^{3+} and Mn^{3+}).

2. WHEN TO USE WHICH RADIATION MODEL?
   It is clear that for a given glass type, if a grid with a very fine cell size can be afforded, the DOM will in most cases produce more accurate results. A drawback of the DOM is that it may show poor convergence and a high grid size dependency if a high absorption coefficient is combined with large cell sizes. In this case, most of the incoming radiative energy that enters a cell is absorbed and transferred to heat, yielding a steep step profile in the main variable of the DOM, the radiative intensity. Also, the large numbers of ordinates that need to be taken into account make the method computationally quite expensive. This is especially true for the spectral DOM, where a transport equation is solved for each spectral band and each ordinate, on top of all the other transport equations for flow, temperature, turbulence, chemistry etc.

   Currently, practically used grids for 3D simulations of complete melting furnaces including melt, refractory and combustion space, typically have 500,000 to 2,000,000 cells. Due to the large number of unknowns per cell, and the equations that often show strong non-linearities, calculation times of a few days are quite common, even for well parallelized programs. This necessitates a trade-off between accuracy (smaller thus more cells) and acceptable calculation times (less cells, simpler and/or less models applied). The choice between Rosseland and DOM is an important factor in this trade-off.

The following rules of thumb can be formulated for the <u>Rosseland Approximation</u>
- The glass domain can be assumed to be <u>optically thick</u>
- Complete absorption of radiation assumed over certain depth H of glass bath
- Rosseland cannot be applied if 10% or more of the radiation from the combustion space reaches the melt bottom
- This leads to the following restriction for application of Rosseland:

$$\kappa \cdot H < -\ln(0.1) \qquad \text{or } \kappa \cdot H < 2.3 \qquad (2)$$

Usage of the <u>DOM</u> is limited by
- A typical glass control volume is required to be <u>optically thin</u>
- When the absorption or emission of radiation within one grid cell is larger than ~ 50% of the incoming radiation intensity, DOM starts to become inaccurate, leading to the restriction:

$$\kappa \cdot \Delta Y < 0.5 \qquad (3)$$

where ΔY is the typical height of a grid cell.

3. A HYBRID ROSSELAND & SPECTRAL DOM APPROACH

The mentioned considerations have led to the formulation and development of a hybrid model that combines the benefits of both methods. It applies the Rosseland approximation whenever justified, and the DOM whenever needed.

Fig. 1 show the spectral absorption of a flint glass in the region where most thermal radiation takes place, between from 1–4 µm, measured at temperatures of 20 °C, 1100 °C, 1200 °C and 1300 °C. It can be seen that at elevated temperatures, the temperature dependency is not very large, and that between 1.5 and 2.7 µm the glass is very transparent. This curve does not show the absorption below $\lambda = 1$ µm; in fact, this glass type is quite opaque at wavelengths below $\lambda = 0.7$ µm.

For such a glass type , first the spectral absorption profile is approximated by a step-wise profile,

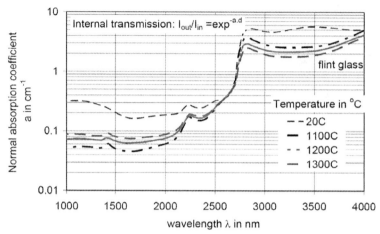

Figure 1. Typical spectral absorption curves of a flint glass at 1100, 1200 and 1300 °C.

where each step represents an absorption band. For each band, it is determined whether the radiative transfer within this band can safely be included in the Rosseland absorption, or whether it should be modeled by the DOM, which then adds one band to the spectral DOM model. The discretized absorption curves of the three low-iron glasses of the next section are depicted in Fig. 2. Note that the temperature dependency within the bands has been dropped for simplicity. In this example, for ultra-clear glass the spectral DOM will be used for two bands and the rest will be modeled via Rosseland. For flint glass, the whole absorption spectrum can be modeled by Rosseland and no DOM needs to be applied. For each spectral DOM band, appropriate weights w_j are used to scale the radiative intensity within that band

$$w_j = \frac{\pi}{\sigma T^4} \cdot \int_{\lambda_{lower}}^{\lambda_{upper}} E_{b\lambda} \, d\lambda \qquad E_{b\lambda} = \frac{2\pi h c_0^2}{n^2 \lambda^5 \left[e^{hc_0/n\lambda kT} - 1 \right]} \qquad (4)$$

Here, $E_{b\lambda}$ is the black-body emissive power (Planck's law).

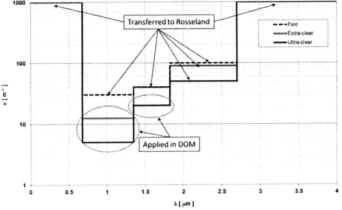

Figure 2. Discretized absorption coefficients for the three glass types.

4. DEMONSTRATION ON AN INDUSTRIAL CASE

In this section, the influence of the applied radiation model of the melt domain on important variables for the glass quality, such as the residence time distribution and the melt temperatures is investigated. In the oxy-fuel float glass furnace that is shown in Fig. 3, calculations have been performed for two soda-lime glasses of varying transparency:

- flint $Fe_2O_3 = 0.08$ wt% (800 ppm)
- ultra-clear $Fe_2O_3 = 0.008$ wt% (80 ppm)

Two above-mentioned types of radiation approaches for the melt will be applied:

- Rosseland Approximation   for flint and ultra-clear
- Hybrid Rosseland/spectral DOM   for ultra-clear

As mentioned, Fig. 2 shows the discretized step-wise spectral absorption curves that are used for the hybrid model.

Figure 3. The oxy-fuel float furnace

The main characteristics of the furnace are summarized in Table 1.

Table 1. Characteristics of the oxy-fuel furnace.

Fuel type	Oxy/gas-fired
Burners	Flat flame & pipe-in-pipe
Pull	150 ton/day
Firing rate	~ 11 MW (~ 1250 m_n^3/hr)
Melter	Length 23 m, width 7 m
Waist	Length 4.7 m, width 2.8 m
Working-end	Length 7.5 m, width 5.8 m
Melting energy	$7.76 \cdot 10^5$ J/kg
Cullet fraction	25%

The only variables that are varied are thus the modeling of the radiative energy transfer in the melt, and the properties of the glass. The <u>combustion space domain</u> is simulated by

- Gas flow and enthalpy transport
- 1D refractory model; this assumes that any heat flowing through the crown and refractory side-walls of the combustion space is always perpendicular to the wall surface. This sub-model allows for less 3D control volumes, speeding up the calculation.
- Mixture-fraction combustion model with two-step reaction mechanism
- Oxy-fuel combustion accounting for intermediate radicals
- Radiation Modeling: Gray DOM model with κ from Hottel charts
- k-ε turbulence model with wall-functions

whereas the following sub-models are applied on the <u>glass melt domain</u>

- Buoyant melt flow and temperature transport
- 3D refractory model
- Batch melting model including latent heat release
- **Radiation Modeling: Rosseland Approximation or Hybrid model**
- Bubbling model (for a bubbling row halfway the melting section)
- Cooled skim bar model (for a skim bar in the waist)
- Cooled vertical stirrer model (for a stirrer row in the waist)

For the ultra-clear glass, mid-plane temperatures from the Rosseland and the hybrid cases are shown in Fig. 4. The main difference is the temperature jump between the melter bottom temperature and working end bottom temperature. When applying the Rosseland Approximation, most radiative heat from combustion space is absorbed near the glass surface, which leads to a higher melt surface temperature and a lower bottom temperature in the melting region than for the results of the hybrid/Spectral model. The same difference still prevails in the conditioning zone after the waist, but is much less pronounced there.

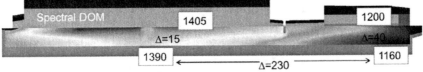

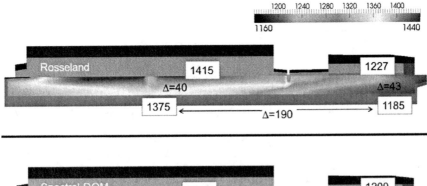

Figure 4. Calculated melt temperatures for the ultra-clear glass. The Rosseland Approximation (top) vs. the hybrid model (bottom).

The higher melt surface temperatures of the Rosseland case lead to a more stable stratified density profile since lighter fluid above heavier fluid is inheritantly stable. Thus the driving force due to buoyancy is lower in the Rosseland case, leading to significantly weaker convection loops. This is shown in Fig. 5 for the velocities beneath the hot-spot, where a difference of about 25% is seen. The maximum velocity differences of up to 75% are observed beneath the batch.

This stronger convection loops that are calculated with the hybrid model are not only quantitative in nature, but they also lead to a short-cut flow. Such short-cuts are mostly hard to detect from velocity plots or animations. But the tracing of many passive particles on the calculated velocities through the melt, and the succesive plotting of the residence time distribution, can reveal such flow paths, that may carry badly molten or

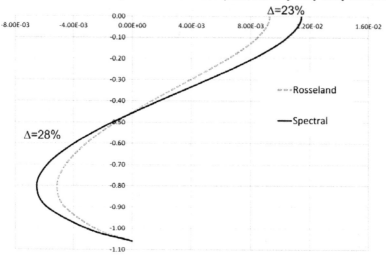

Figure 5. Profiles of the horizontal velocity component on a vertical line in the fining area for the ultra-clear glass. The Rosseland approach (gray) vs. the hybrid model (black).

unrefined glass volumes to the outlet, possibly without having seen any high temperature at the melt surface. Fig. 6 shows the residence time distributions for the ultra-clear glass. The high peak at low residence times, for the hybrid/spectral DOM calculations, is a sign of a short-cut flow. It is not present for the Rosseland results, and would thus not have been detected when only this simplest of the radiative models would have been used.

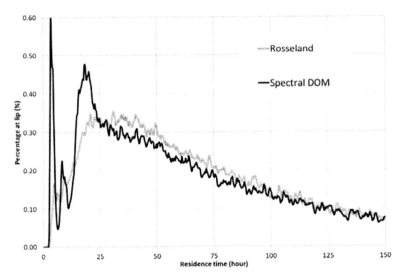

Figure 6. The residence time distribution for the ultra-clear glass. The Rosseland Approach (gray) vs. the hybrid/spectral DOM model (black).

Table 2 sums up the main temperature values for all calculated cases, showing both the significant influence of the glass transparency, as the differences caused by the applied radiation model for the melt.

Table 3 shows a comparison of the calculation times. As can be seen, applying the Rosseland Approximation clearly is cheapest, but the additional effort for the gray or hybrid/spectral DOM remains within reasonable bounds of 50%. All calculations were performed on the same grid – this was also the assumption for the stated criteria for the applicability of the radiation models.

Table 2. Key temperature results of the test cases.

		STANDARD FLINT Fe = 0.08 wt%	ULTRA-CLEAR Fe = 0.0008 wt%	
		Rosseland Approximation	Rosseland Approximation	Spectral Hybrid DOM
Melter crown Hot Spot Temperature	[°C]	1559	1506	1497
Melter bottom Hot Spot Temperature	[°C]	1311	1375	1390
Glass surface Hot Spot Temperature	[°C]	1546	1472	1466
Average Lip Temperature	[°C]	1200	1188	1163
Backflow from waist to fining area		392%	704%	763%

Table 3. Comparison of calculation times.

	# bands	# processors	CPU Time [Hr]	# iters	CPU/Iter [sec]
Rosseland	0	8	12.5	4500	80
Gray	0	12	18.2	4467	176
Spectral Hybrid	1	12	17	5121	143
Spectral Hybrid	2	12	16.1	3947	176

5. CONCLUSIONS

The validity of the Rosseland Approximation, and the required number of grid cells for a normal or spectral DOM model, necessitates a flexible approach for radiation modeling in glass melts, especially for ultra-clear glasses and shallow feeders. The presence of opaque bands ($\kappa \geq 100$ m^{-1}) in the absorption spectra has led to the formulation and development of a hybrid option for spectral DOM, which avoids both excessive CPU-times and decreased accuracy. With this extension, GTM-X now offers the following flexibility:

- Rosseland model
- Gray DOM for glass and/or combustion space
- Spectral DOM for glass and/or combustion space
- Hybrid Rosseland/Spectral DOM approach for glass

The modeling results for an industrial oxy-fuel float furnace show that for regular flint glass, the Rosseland Approximation is valid, but for glasses with a lower iron content, a hybrid Spectral Radiation Model is required in order to correctly predict:

1. (Max.) bottom temperatures (structural integrity / corrosion / bubbles)
2. (Max.) glass temperature (fining)
3. Recirculation flows (residence time / quality)

6. LITERATURE REFERENCES

[1] Rosseland, S. Note on the absorption of radiation within a star. M.N.R.A.S., 1924, 84, 525
[2] Ferziger, J. H. & Perić M., Computational Methods for Fluid Dynamics, Second Edition. Springer-Verlag, 1999
[3] A.M. Lankhorst, A.J. Faber, Spectral radiation model for simulation of heat transfer in glass melts, Glass Technnol.: Eur. J. Glass Sci. Technol. A, April, 49, 2 (2008) 73–82
[4] A.J. Faber, Optical properties and redox state of silicate glass melts, C.R. Chimie 5 (2002) 1-8

NOVEL METHOD FOR STRESS INSPECTION OF TEMPERED OR THERMALLY STRENGTHENED GLASS

Sarath Tennakoon
Emhart Glass Research Center
123, Great Pond Drive, Windsor, CT, USA

ABSTRACT

Thermal strengthening or tempering of glass has existed for more than 50 years. Due to the lack of an efficient accurate method for online inspection, the quality of the strengthening or tempering is measured on a sample basis using either destructive or non direct laboratory methods. Emhart Glass Inc. has developed a new optical inspection method that can be used to inspect the profile of the stress distribution non-destructively and obtain the quality of thermally hardened/tempered glass samples faster. The detail of the method, measurement data of various glass samples, and the limitations of this method are presented.

INTRODUCTION

Thermal strengthening or tempering has been used to increase the durability of glass for several decades. Strengthened or tempered glass is mechanically and thermally much stronger than regular glass and less susceptible to damage from mechanical loads or thermal shock. Thermal strengthening is achieved by fast quenching the heated glass samples above the annealing point to below the glass strain point. This process creates compressive stresses on the surfaces of the glass balanced by tensile stresses in the middle of the glass wall. During the manufacturing process the quality of the strengthening or tempering is measured on a sample basis using either destructive or non direct laboratory methods available in the market today. Quality of the tempering is maintained by controlling the thermal strengthening process during the production runs. There are several non-destructive glass stress inspection methods in the market today. Most of these methods use the visible polarized light and the birefringence properties of the glass under stress to measure the levels and quality of the stress distribution within the glass. Some of those are (a) Imaging polariscope, (b) Immersion polarimeter, (c) Grazing angle polarimeter, and (d) Scattered light polariscope.

The imaging polariscope uses polarized white light and a polarization analyzer to measure the net retardation of the light as it passes through the glass piece (http://www.agrintl.com/products/view/18/Polariscope). It measures the residual stress or strain across the glass to assess the quality of the annealing of the glass samples after production[1]. This apparatus cannot measure the stress distribution inside the glass wall that is required to obtain the quality of the heat strengthened or tempered glass samples.

The immersion polariscope requires the glass piece be inside an index matching fluid in a non stressed rectangular glass vessel. The immersion polariscope uses a polarized light source and polarization analyzer to measure the stress inside curved glass

259

walls immersed in the index matching fluid[2]. It uses the changes to the polarized light retardation by layers of the curved glass wall section to obtain the average stress distribution inside the glass from the outer wall surface to the inner wall surface of a glass segment (http://www.glasstress.com /ap07.htm).

The third instrument, the grazing angle surface polarimeter (GASP®) uses changes to the interference patterns created by the low angle polarized light beam passing through the glass wall to measure the glass stress levels near the surface of the wall (http://www.strainoptics.com/files/GASP%20Bulletin.pdf)[3]. This system also requires index matching fluid between the instrument and the glass surface and a beam block to prevent light traveling above the surface of the glass, and force the grazing light beam to pass inside the surface of the glass wall.

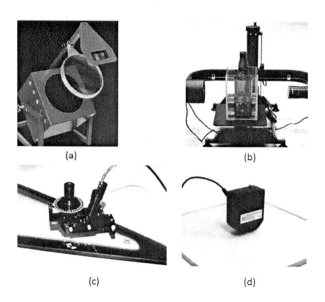

(a)

(b)

(c)

(d)

Figure 1. (a) The Imaging Polariscope, (b) Immersion Polariscope, (c) Grazing angle surface polarimeter, (d) Scattered light polarimeter, SCALP - 04

The scattered light polarimeter uses the changes to the scattered light intensity created by a plane polarized laser beam along its path as it passes through the glass wall (http://www.glasstress.com/scalp03.htm). It uses multiple orientations of the polarization plane of the laser beam around its axis and captures multiple images of scattered light intensity patterns to obtain the stress distribution along the laser path (or across the glass wall)[4]. All the methods that can measure the stress distribution at or near the surface or inside the glass require index matching liquid between the instrument and the glass surface and will not be practical instruments to use in a production environment to measure the stress distributions inside glass walls.

The method described in this paper is relatively simple to use in a production environment and does not require index matching fluids. It uses scattered fluorescent light created along its path by a plane polarized laser beam passing through the glass wall. This instrument can measure the stress across a thickness of a flat or curved glass segment and obtain both the thickness of the glass as well as the thicknesses of the stress layers inside the wall.

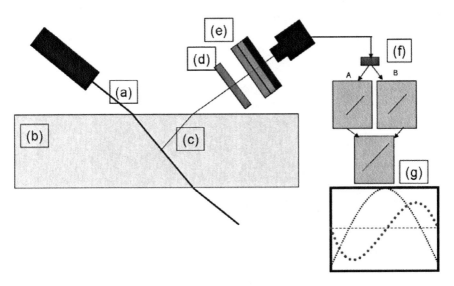

Figure 2. Components of the fluorescence light polarimeter

The inspection system consists of several components (Figure 2.): (a) A polarized laser beam that enters the glass wall at a certain orientation, (b) a test specimen (c) a partially polarized fluorescence light produced by the laser inside the glass wall, (d) a band pass filter to filter fluorescence light, (e) a polarization analysis system to separate the polarization into two orthogonal components, (f) an imaging system to obtain the two polarization component images, and (g) image analysis system to analyze the images and obtain the stress field.

A plane polarized focused laser beam shines on to the glass surface at a certain orientation and refracts into the glass wall. Angles of the incident of the laser beam are chosen for the coupling mediums that the laser travels through before entering the glass to maximize the signal level obtained by the inspection system. The linearly polarized laser beam enters the wall of the glass from the outer surface and is refracted in to the glass. As the polarized laser beam travels through the glass it produces scattering along its path as well as excites the electronic state of some of the elements and produces the fluorescence (Figure 3.).

(a) Scattered and fluorescence light along the laser path inside the glass

(b) Fluorescence light after filtering the excitation laser light

Figure 3. Scattered and fluorescence light produced by excitation laser shined on to glass piece

The fluorescence light produced at each point along the laser beam is partially polarized with the largest component perpendicular to the laser polarization plane at that location. The fluorescent light acts like a small partially polarized localized light source inside the glass wall. As the polarized fluorescence light from each of these sources passes through the varying stress field toward the camera their polarization characteristics will change from linear to elliptical to circular to elliptical to linear, the pattern is repeated due to the birefringence property of the glass under stress. That fluorescent polarized light that has passed through the stress field in the glass wall is separated into two linear components using the polarizing analyzer. Then, a CCD camera creates an image of the polarization intensity at a first linear polarization direction and a second linear polarization direction into two camera images. A band pass light filter is used to block the scattered light reaching the imaging system. The intensities on each image represent the polarization components of the fluorescence light as it emerges from the glass wall on to the imaging camera. From two images, the normalized difference image can be obtained. By analyzing the intensity difference at each pixel location along the fluorescence line of the two polarization component images, the stress field can be obtained.

The normalized difference of the data along the fluorescence line on a stressed sample produces a nominally rotated S-shaped retardance curve. By curve fitting a polynomial to that data and taking the derivative of that curve one can obtain a parabola that is representative of the stress field across the thickness of the wall.

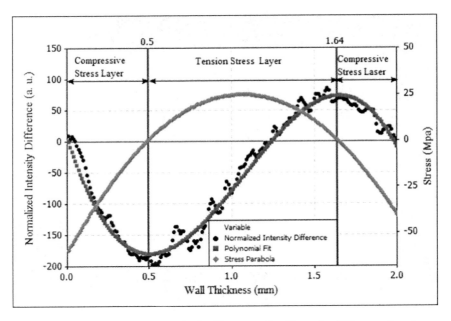

Figure 4. The Retardance curve obtained by normalized intensity difference from two polarization component images.

When there is no stress, no polarization changes occur in the fluorescence light and no intensity difference between the normalized images and the normalized intensity difference will be zero along the fluorescence curve. Figure 4 shows the retardance data (black dots) taken at one point on a thermally strengthened glass piece. The S-curve is a 3rd order polynomial fit to the retardance data and the parabolic curve is the differentiation of the polynomial fit, which represent the stress field across the thickness of the wall. The stress parabola not only gives the stress distribution across the wall, it also gives the compressive and tension layers thicknesses as well as the wall thickness of the glass wall. Once calibrated, the stress parabola indicates both type and magnitude of the stress within the glass wall, with compression being indicated by negative values and the tension being indicated by positive values.

Emhart glass has developed two measurement systems (heads) using this technology. One is optimized to work with water as the coupling medium between the head and the glass wall. The other system is optimized to work with air as the coupling medium.

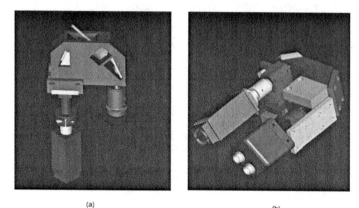

(a)

(b)

Figure 5. Measurement heads, (a) optimized for water as the coupling medium and (b) optimized for air as the coupling medium

The glass stress measurement system and method is also applicable to measuring stress distribution across thermally strengthened (tempered) flat glass or curved glass segments, and is capable of accurately measuring stress in and the thickness of the walls. This system is capable of high speed measurement of the stress, since it requires only two images to obtain the stress at a single location. The system is adaptable to large scale production of thermally strengthened flat glass or curved glass segments manufacturing or glass container manufacturing.

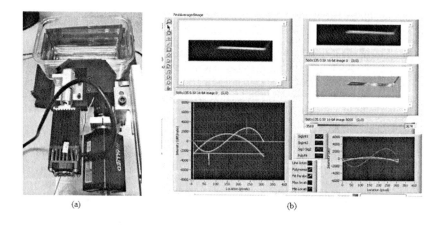

(a)

(b)

Figure 6. (a) Measurement of stress on a sidewall of a storage container, (b) images of fluorescence components and the retardance and stress plots

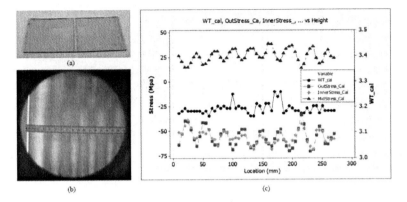

Figure 7. (a) Tempered auto glass panel. (b) Imaging polariscope image of the panel showing the net stress variation of the panel across panel due to the cooling jets motion. (c) the stress measurement data across the panel using Florescence Light Measurement system.

LIMITATIONS

Fluorescence light stress measurement system requires fluorescence created by polarized laser light beams to be partially polarized. Not all glasses produce polarized fluorescence when excited with a polarized laser beam. Most of the container glasses contain minute amounts of fluorescence producing elements due to the impurities in the batch material used or additives used to change the color of the glass. For other non-fluorescing glasses, a minute amount of additive (most probably one or more rare earth element) need to be added to the glass to produce the fluorescence. Also, stress on some of the colored glasses may not be measured using this method due to the absorption of light from the excitation laser and/or fluorescence light by the glass.

SUMMARY

Fluorescent light produce by a polarized light beam inside a glass wall and birefringence properties of the glass under stress can be used to accurately measure stress across the wall. The stress measurement system and method developed using the technology presented in this paper is relatively simple to use. It can be used to obtain stress across a glass wall of thermally strengthened flat, curved or container glass segments. The system is capable of measuring wall thickness and the thickness of each of the stress layers. The measurement heads can be used with or without index matching fluids and measures at a high rate. It is suitable for online inspection of thermally strengthened glasses.

REFERENCES

[1] AGR international, Inc, *Evaluation of residual strain in rigid containers*, Polariscope_redesign.pdf

[2] Hillar Aben, Andrei Errapart, and Leo Ainola, *On real and imaginary algorithms of optical tensor field tomography*, Proc. Estonian Acad. Sci. Phys. Math., 2006, 55, 2, 112–127.

[3] GASP® Instruments, Bulletin GSP-0112, www.strainoptics.com.

[4]. H. Aben, L Ainola, J Anton, *Integrated photoelasticity for nondestructive residual stress measurement in glass*, Optics and Lasers in Engineering, Vol 33, 1, 2000, 49-64.

[5]. U. S. Patent 8,049,871 B2, "Glass Stress Measurement Using Fluorescence".

Author Index